AF885226

It's A Variable World
An Introduction to the Art of Algebra

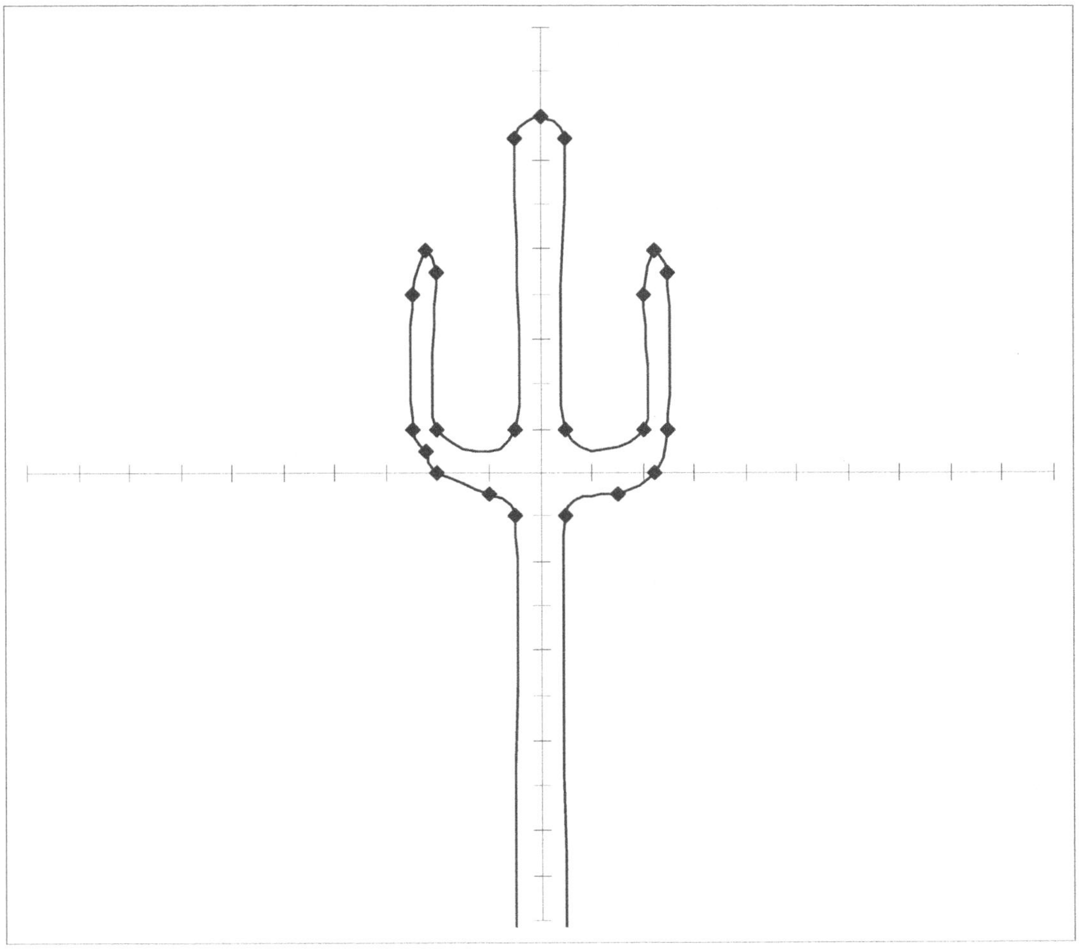

Written and illustrated by
Joan Schoepfer

Mathematics is, above all, a language we use to communicate with one another.

Tips for getting the most out of this book

To learn a language requires active engagement.

1. Read (and understand) every step in every example. (They are not nearly as formidable line by line.)
2. Important concepts are in frames for easy reference.
3. Talk with your classmates, teachers, friends, and others about these concepts.
4. Work all of the problems. They are few and offer valuable insights.

HAVE FUN!
(first important concept)

Acknowledgements

Through the years I have been blessed with helping to guide many students in mathematics. Through their struggles and successes these students have helped me to understand the blockades and obstacles that we, as a society and as educators, have placed in the path to their understanding of this crucial subject. To name them all would fill this volume, but each has left their own unique mark on this work.

My very first specific acknowledgement must go to my sons whose struggles and successes compelled me to look beyond the norm for teaching techniques.

I will always be grateful to Mark at UNM Gallup for allowing me to teach multiplication of fractions before addition, my first venture into changing the status quo for mathematics education.

Tutoring at the Zuni Adult Learning (ADOBE) Center in Zuni, New Mexico was truly the seed for the method presented in this book. The warm hospitality and culture there, openness to experimentation with new methods, coupled with the avid dedication of many of the students helped me introduce, evaluate, and fine tune the basic concepts that have grown into this book.

I wish to thank my granddaughter who graciously shared my attention with countless students, jobs, and tasks so that I could complete this work. Through her needs, desires, and antics she helped me solidify the concepts used in this book. She also led me to the Village Home Educational Resource Center in Beaverton, Oregon. This amazing place and the people that are the heart of it took the work to a whole new level. It was here that I began to incorporate the connection between art and mathematics and other higher level attributes to the method. It was also here that I found young fresh minds, some completely unspoiled by prior math education. The concepts for the basics method flourished and grew under this nourishing environment. These basic concepts are at the root of It's a Variable World. They are what make it different, effective, and powerful. The Village Home Educational Resource Center will retain a place in my heart for many reasons, but here I wish to express mostly my most sincere gratitude for their contributions to this book.

I would be remiss not to mention the small group of homeschoolers, the Barnraisers, in Eugene, Oregon who gave up so many beautiful play days to come, count toothpicks, and draw pictures of math relationships (and other things) with me. This book, and certainly the next, would not be what it is without you.

It was the Tillamook Bay Community College that gave me the final puzzle piece, the opportunity to try the "book" on a group of innocent college students. I will be forever grateful to the school, and those students, for proving that it works.

I am eternally grateful to all of you. Thank you.

Table of Contents

Chapter 0 – What is a variable and why do I care?

0.1 What is a constant?

In prior books and lessons we have learned to use numbers to keep track of, understand, and explore our world. We have seen how the development of our number system and many shortcut operations has made this easier. We have used numbers to represent many specific quantities at specific points in time. Although we use the same numbers to represent many things, once we assign a number to the quantity and time in question, it never changes. It always remains the same. In the language of mathematics, we call these numbers constants. Examples of these constants are everywhere. A few are given below:

12 = the number of eggs in a dozen
12 = the number of signs in the Zodiac
5 = the number of children in the Jones' house on May 1, 1970
3 = the number of owls at Snowy Point on March 2, 2002

Name that Constant Game

One team or student states a number.
Within a specific time frame the second team or student
must come up with a given number of constant quantities
that can be represented by that number.
The team with the most giggles wins.

Note: Everyday life, trivia, science, virtual reality, and
many other resources many be drawn on for answers.

0.2 What is a variable?

As we begin to explore that world, however, we note that there are many things of interest that can not be represented by a specific quantity at a specific time. For instance we may be concerned with the population growth or decline of owls, or how different housing types accommodate different family compositions. We keep track of these changes by letting a letter or symbol take the place of the number for a quantity that is changing, for instance:

x = the number of children in each house on Pine Street on December 1, 2000 or,
N = the number of owls at Snowy Point each year (always counted on March 2)

The operations we developed for numbers were based on logic. Once we identify a variable we can use it in all the same operations using the same logic as we did with numbers. As you will see as you work through this book, the ability to do this is a powerful tool in mathematics. Using variables in mathematical expressions and operations allows us to accomplish three major tasks. They are:

1. Making General Statements (or "Stating the Obvious")
2. Exploring Relationships
3. Solving for Unknowns

Unit 1 – Stating the Obvious

The use of variables allows us to turn specific mathematical problems into general rules. These rules can then be used to rapidly help us complete complicated tasks.

Monkey Hanging From a Tree ---

Chapter 1 – Review of the Basics (in a new light)

Each person studying this book has had a general preparation in arithmetic. In this chapter we will review the rules and concepts of arithmetic. We will then use variables to write general statements that express these rules. The general statements will then be used to help us simplify and evaluate complicated mathematical problems and expressions.

1.1 Addition and Subtraction

The most basic mathematical operation, addition is just a shortcut for counting. When I add my 15 monkeys to your 42 we don't have to count them to know there are now 57. We can use memorized facts and logical methods to find the answer much quicker and easier than trying to count all those monkeys.

The symbol for addition is +. To state mathematically that we are adding 2 things to 3 things we write the expression

$$2 + 3.$$

Since we know the sum of 2 + 3 (or we can easily count it), we can use another mathematical symbol, =, to express that when 2 things and 3 things (called addends) are added together we get 5 things (called the sum).

$$2 + 3 = 5.$$

1.1.1 Identity Property of Addition

We know that if we add zero (nothing) to five we will still have 5 of whatever we started with. We can show this relationship is true for many different constants in mathematical sentences.

$$0 + 5 = 5; \quad 35 + 0 = 35; \quad 14 + 0 = 14$$

It is obviously true for any number, but to write all of the possibilities would be impossible. It is here we will make our first general statement through the use of variables. First, we must make a statement defining the variable we wish to use so that people who read the statement can understand what we are trying to say. In this case, a broad definition will serve us best. Once we define a, we can make a statement like we did before substituting a for the constant:

> Let a = any real number, then
> $$0 + a = a$$
> Identity Property of Addition

Now we have stated that when we add zero to any number we will get back the number itself (without having to write the statement an infinite number of times!). Any real number anyway (although this one works for imaginary ones too).

1.1.2 The Inverse Operation of Addition (Subtraction)

We undo addition with subtraction or, stated another way; subtraction is the inverse operation of addition. If I take that group of 5 things and remove 3 of them I will be left with my original 2. I can express this using the symbol for subtraction (-) as

$$5 - 3 = 2.$$

or, more specifically,

$$2 + 3 - 3 = 2.$$

The last statement says that if I have 2 things and add three things to them but then subtract 3 things back out, I will be left with the original two. This is a very simple, and obvious, statement but a very important one as we shall see in Unit 2.

If we want to show that this relationship is true for any numbers (which it obviously will be), we can use variables instead of numbers. First, we must define the variables (this time we have two).

Let a = any real number
Let b = any real number

Then we can state that

$$a + b - b = a$$

Inverse Operation of Addition (Subtraction)

The identity property of addition also applies to subtraction:

$$0 - a = a$$

1.1.3 Commutative Property for Addition and Subtraction

One of the first things we notice when we start adding things together is that it really doesn't matter what order we add them. Whether I add my 15 monkeys to your 42 or you add your 42 monkeys to my 15 we will still end up with 57 monkeys. To show this, we can commute, or move around, the numbers any way we want in our mathematical expression. For the monkeys it looks like this:

$$15 + 42 = 42 + 15 = 57.$$

If we want to say this is true for any numbers, we can again use variables. First, we must define the variables.

Then we can state that for a and b equal to any real number,

$$a + b = b + a$$

Commutative Property for Addition

It is not hard to see that order is also unimportant when removing things. What we do have to be careful to watch, however, is whether we are adding or subtracting. For this reason, it is important to commute the operation sign when commuting numbers or variables being added or subtracted. We can write this expanded rule as shown below. [Note: we often omit an operation sign at the beginning of a sentence.]

Let a = any real number
Let b = any real number

$$a - b = - b + a$$

Commutative Property for Addition and Subtraction

Example

Sam owed me 7 dollars and then paid me back 10. I want to show this as a mathematical statement and use it to find out what Sam still owes me.

First I can write a mathematical statement to represent the situation. Sam removed 7 dollars from me (he owed it to me), then he added 10.

$$- 7 + 10$$

When I look at this expression it is hard to see the answer, but using the commutative property for addition makes it more obvious.

$$- 7 + 10 = 10 – 7$$

$$10 – 7 = 3$$

It looks like Sam gave me a little too much. I now owe Sam $3.

1.1.4 Puzzling It All Out

Learning to play a musical instrument, build a canoe, or any other worthwhile endeavor requires active involvement before the information provided can be fully integrated. The study of mathematics is no different.

Work each of these puzzles on your own until you are sure you have the correct answer. Then check the answer in the back of the book. If you are having trouble agreeing with the book's answer you can: check the solutions manual if one is available in your area or ask an online program to show you a solution, ask a friend or a tutor, or ask your teacher for advice. Keep in mind that there may be more than one way to solve each problem. Always look for the easiest solution.

It is preferable that you understand how to get the correct solution for every puzzle on this page before moving to the next section.

Evaluate:

1. $5 + 6 + 14 + 25 =$

2. $17 + 8 - 0 - 7 - 4 =$

3. $20 - a - 18 =$

4. $-7 + q + 10 - q =$

Write:

5. A mathematical statement that shows the sum of Harry's and Judy's savings. Make sure you define your variables.

6. A general statement stating that the perimeter (P) of a rectangle is always equal to the sum of it's sides (a and b).

1.2 Multiplication and Division

Mathematicians are always looking for easier ways to do things. Although addition and subtraction help us keep track of things, it can be cumbersome and time consuming. If we add the same number of things together repeatedly, a pattern emerges that suggests a more efficient alternative, an operation that counts the number of times the same number is being added. For example, if someone always adds 4 monkeys to our group every month and I want to know how many monkeys he has added after 3 months I have a few choices:

month 1

month 2

month 3

1. I can count the monkeys (May be tough since they're already in the group.)
2. I can add

$$4 + 4 + 4 = 12.$$

3. I can .... MULTIPLY!

I just make a new operation – a new statement – a new symbol.
(Isn't mathematics resourceful?)

The symbol for multiplication is to write two numbers (or variables) next to each other but we must be careful when we do, especially with numbers. We often use a parenthesis ( ), to help us clarify what we mean. We can also use a • or * between the two things being multiplied (factors). Using these symbols we can write the monkey problem out in multiplication like this:

$$4(3) = 4 \bullet 3 = 4 + 4 + 4 = 12$$

[Note: Since x can also be used as a variable, we never use this symbol for multiplication in Algebra.]

We call the numbers (or variables) being multiplied factors and the answer, the product.

1.2.1 Definition of Multiplication

If we want to define the operation of multiplication for all real numbers we can use variables to describe it.

For x = any real number,
$x + x + x + x = 4x$
Definition of Multiplication

Food for Thought
Can we write a truly general definition for multiplication? What would be required?

1.2.2 Zero Product Property

Unlike addition, multiplication by zero does not return the original number. Instead, as we all know, it always results in a product of zero,

$$1(0) = 0; \quad -4(0) = 0; \quad 0(367) = 0.$$

Using variables we can make a general statement about this

If a = any real number

Then $a(0) = 0$

and

If $ab = 0$

Then either $a = 0$ or $b = 0$.

Zero Product Property

1.2.3 Identity Property for Multiplication

In multiplication, we get back our original number by multiplying by one,

$1(4) = 4,\ 18(1) = 18$ so we can state,

If a = any real number

$$a(1) = a$$

Identity Property for Multiplication

1.2.4 Inverse Operation of Multiplication (Division)

The inverse operation of multiplication is division. Although we can think of division as repetitive subtraction, it is often easier just to think of it as undoing multiplication. The symbol for division is a line placed beneath the number representing the original quantity (the dividend). The number of subtractions (the inverse of the number of multiplications or the divisor) is then written under the line like this:

$$\frac{12}{4} = 3.$$

(read 12 divided by 4 equals 3)

[Note: When working only with numbers some people use the symbol ÷ for division. Although this symbol can be useful in separating complex expressions, it often creates confusion especially in combined problems. The use of this symbol except where absolutely necessary is strongly discouraged.]

We can state the definition of division as the inverse of multiplication as shown:

Let a and b equal any real numbers,

$$\frac{ab}{b} = a$$

Inverse Operation of Multiplication (Division)

Both the Zero Product Property and the Identity Property for Multiplication apply to the inverse operation (Division) as well. Often, the identity property for division is convenient for keeping track of mixed multiplication and division problems.

Example

Multiply $\frac{2}{3}$ by 6.

When we write this in a mathematical expression it may not be clear which things are being multiplied and which divided.

$$\frac{2}{3}(6) =$$

If I remember $\frac{6}{1} = 6$, I can easily make the expression clearer through substitution.

$$\frac{2}{3}\left(\frac{6}{1}\right) =$$

I can then see that I am simply need to multiply 2 by 6 and divide the result by 3 obtaining

$$\frac{2}{3}\left(\frac{6}{1}\right) = \frac{12}{3} = 4.$$

1.2.5 Rational Expressions

Any division problem can also be expressed as a ratio. A ratio expresses the proportion of one thing to another. For example, the proportion of ethanol in most commercial gasoline could be written as 1 part ethanol to 10 parts gasoline, or 1:10. The corresponding division problem, $\frac{1}{10}$ tells us the same thing. Either way you look at it the ethanol is still 10% of the gasoline. Because of this fact, we have come to call the symbolism for division rational form and expressions that contain division forms are called rational expressions.

In our recent analysis, we ran into several rational expressions that, if evaluated, resulted in decimals. In some cases those decimals were infinitely repeating or even irrational.

Often, in Algebra, we leave answers expressed as rational expressions rather than expressing them as decimals. If the expression is given in its lowest form it provides a concise, finite answer that a rounded decimal may easily be extracted from if desired.

1.2.6 Commutative Property for Multiplication and Division

If we look again at a picture of the monkeys coming, 3 in a group 4 times, it is again obvious that it does not matter whether we have 4 groups of 3 monkeys or 3 groups of 4 monkeys. We will have 12 monkeys in either case.

$$3(4) = 4(3)$$

Letting a and b equal any real number, we can state

$$ab = ba$$

Commutative Property for Multiplication

Although not as obvious, Division can also be done in any order. For example, the expression we obtained in 1.2.4 is, indeed, much easier to decipher, but is not readily analyzable for the first step, dividing 2 by 3, yields an infinitely repeating decimal.

$$\frac{2}{3}\left(\frac{6}{1}\right) = ?$$

Using the commutative property for multiplication and division, we can instead first divide 2 by 1, then 6 by 3. Multiplying yields a simple and finite solution.

$$\frac{2}{1}\left(\frac{6}{3}\right)=2(2)=4$$

As with Addition and Subtraction, we must always remember to keep track of whether we are dividing or multiplying. We can write this property as:

For a, b, e, and f equal to any real numbers,

$$\frac{a}{e}\left(\frac{c}{f}\right)=\frac{a}{f}\left(\frac{c}{c}\right)$$

Commutative Property for Division

As with addition, we can use the commutative property to help us quickly resolve complex problems.

Example

$$\textit{Multiply } \frac{7}{4} \textit{ by } \frac{2}{7}$$

This is a difficult task to do directly as both expressions result in irrational numbers (decimals that just go on forever).

However, commuting the divisors significantly simplifies the task

$$\frac{7}{4}\cdot\frac{8}{7}=\frac{7}{7}\cdot\frac{8}{4}=1(2)=2\,.$$

1.2.7 Multiplication and Division of "Fractions"

In the past, many educators have presented methods for dealing with rational forms as rules for "Fractions". Although fractions are actually parts of a whole and can be expressed as decimals and percentages as well as rational forms, the practice of substituting the word fraction for rational form is quite common. Although a separate set of rules is often taught for rational forms (fractions), all of these rules and methods have their roots in the basic whole number operation of division and how it interacts with other operations.

Many of you may have seen the commutative property discussed in the previous section as the cross canceling method for multiplication of fractions. Indeed, the divisors do not need to be physically moved to apply the principle. They can be divided diagonally. The resulting method is shown below.

$$\frac{\overset{1}{\cancel{7}}}{\underset{1}{\cancel{4}}}\cdot\frac{\overset{2}{\cancel{8}}}{\underset{1}{\cancel{7}}}=\frac{1(2)}{1(1)}=2$$

Although the cross canceling method was developed for fractions it is really just a result of the commutative property for multiplication and division. As seen above, it is as applicable to whole numbers as it is to fractions.

[Note - diagonal division (cross canceling) works but can be difficult to keep track of. Be careful to clearly show what you do in case you have to check it later.
If you are uncertain or getting questionable results, commute the divisors instead to minimize errors.]

1.2.8 Complex Rational Forms (Division of "Fractions")

Another method taught under the rules for "fractions" and commonly called "division of fractions" also stems from a property of division that develops when we have rational forms within rational forms such as

$$\frac{\left(\frac{a}{b}\right)}{\left(\frac{c}{f}\right)}.$$

To understand these forms it will help to look at very simple numeric examples. We will start with considering what we will get if we try and take one apple and divide it into groups that are one half apple each. We can express this mathematically as:

$$\frac{1\cdot apple}{\frac{1}{2}\cdot apple}.$$

Obviously if we do this we will get 2 groups with one half apple each so we can easily state our observation as

$$\frac{1 \cdot apple}{\frac{1}{2} \cdot apple} = 2.$$

Remembering the identity property for division, we can also express this as

$$\frac{1}{\frac{1}{2}} = \frac{2}{1}.$$

We might notice that the answer is the inverse of the divisor. This would seem intuitive since multiplication is the inverse of division, but we cannot make a generalization on one example. We can, however, observe that this is true no matter how big we want the apple slices. If we want slices sized $\frac{1}{3}$ apple each we will need $\frac{3}{1}$ groups, if we want slices sized $\frac{1}{20}$ apple each we will need $\frac{20}{1}$ groups. Furthermore, we can see that this is true no matter how we represent the size of the slices. For instance, a slice sized $\frac{1}{3}$ could also be represented by a size of $\frac{3}{9}$ and we can easily see that

$$\frac{1}{\frac{1}{3}} = \frac{1}{\frac{3}{9}} = \frac{9}{3} = 3,$$ which we know to be true.

This shows that we will always get the inverse of the divisor when we divide by a rational form. Although there are more formal proofs available, we will let this suffice and state that

For a, b, c, and f equal to any rational numbers,

$$\frac{\left(\frac{a}{b}\right)}{\left(\frac{c}{f}\right)} = \left(\frac{a}{b}\right)\left(\frac{f}{c}\right).$$

Inversion of Divisor Property for Division by Rational Forms
(or Division of "Fractions")

When rational forms appear in the dividend, it is easy to see that we are really just doing successive division. Again using constants and following the order of operations we find that

$$\frac{\left(\frac{18}{3}\right)}{(3)} = \left(\frac{6}{3}\right) = 2\,.$$

If we write the divided by 3 as multiplied by $\frac{1}{3}$ (based on the previous discussion), we can rewrite the expression as below revealing that this solution is also based on the inverse of the divisor property shown above.

$$\frac{\left(\frac{18}{3}\right)}{(3)} = \frac{1}{3}\left(\frac{18}{3}\right).$$

Once we have reformed complex rational forms into simple rational forms we can use any properties available for multiplication and division to obtain an answer. In this case, we will use the commutative property and multiply the divisors first giving

$$\frac{\left(\frac{18}{3}\right)}{(3)} = \frac{1}{3}\left(\frac{18}{3}\right) = \left(\frac{18}{9}\right) = 2.$$

These properties apply to rational forms that yield both fractional and whole number results.

Example

Find the result when $\frac{1}{3}$ *is divided by* $\frac{4}{27}$

First, we write the problem as a mathematical expression,

$$\frac{\frac{1}{3}}{\frac{4}{27}}.$$

Next, we invert the divisor and multiply as shown.

$$\frac{\frac{1}{3}}{\frac{4}{27}} = \left(\frac{1}{3}\right)\left(\frac{27}{4}\right).$$

We can then use the commutative property to obtain the lowest whole number rational form possible:

$$\frac{\frac{1}{3}}{\frac{4}{27}} = \left(\frac{1}{3}\right)\left(\frac{27}{4}\right) = \left(\frac{27}{3}\right)\left(\frac{1}{4}\right) = \frac{9}{4}.$$

Note that dividing nine by four results in a decimal. Although the decimal is finite so the answer can be expressed exactly numerically, we often leave numbers in whole number rational form, avoiding decimals altogether.

1.2.9 Puzzling It All Out

Learning to play a musical instrument, build a canoe, or any other worthwhile endeavor requires active involvement before the information provided can be fully integrated. The study of mathematics is no different.

Work each of these puzzles on your own until you are sure you have the correct answer. Then check the answer in the back of the book. If you are having trouble agreeing with the book's answer you can: check the solutions manual if one is available in your area or ask an online program to show you a solution, ask a friend or a tutor, or ask your teacher for advice. Keep in mind that there may be more than one way to solve each problem. Always look for the easiest solution.

It is preferable that you understand how to get the correct solution for every puzzle on this page before moving to the next section.

Evaluate:

1. $4(6)=$

2. $-3(-2)=$

3. $\dfrac{\frac{a}{2}\cdot\frac{2}{ab}}{\frac{1}{b}}=$

4. $-\frac{12}{3}(-2)=$

Write:

5. A mathematical statement that shows the total amount in my piggy bank if I put \$5 a week of my allowance in for any given number of weeks. Make sure you define your variables.

6. A general statement stating that the area (A) of a rectangle is always equal to the product of it's two sides (a and b).

1.3 Exponents

Once we developed multiplication, we noticed that sometimes we had repetitive problems with identifiable patterns just like in addition and therefore, of course, sought a way to exploit this to make calculations easier (and much less boring!). Repetitive multiplication of the same number is one pattern that is easy to identify and tedious to perform.

$$3(3)(3)(3) = 9(3)(3) = 27(3) = 81$$

$$2(2)(2)(2)(2)(2)(2) = 4(2)(2)(2)(2)(2) = 8(2)(2)(2)(2)$$
$$= 16(2)(2)(2) = 32(2)2 = 64(2) = 128$$

A new operation and symbolism, exponential format, was defined. In exponential format the base is the number being repeatedly multiplied. It is written along the same level and size as the other factors or addends in the problem. The exponent is the number of times it is being multiplied and is written slightly smaller, elevated, and to the right of the base. Once we have something in this form, we can again use fact memorization to shorten our work. For the problems above this looks like:

$$3(3)(3)(3) = 3^4 = 81$$

$$2(2)(2)(2)(2)(2)(2) = 2^8 = 128$$

1.3.1 Definition of Exponential Format

We can make a more general definition of exponential format using variables:

If a and b are any real numbers

Then $a(a)(a)(a)(a)(a) = a^6$

Definition of Exponential Format

Unfortunately, we cannot write a completely general definition without the use of series format, a topic introduced in more advanced math classes.

1.3.2 Inverse Operation

The inverse operation of an exponent is a root. The symbol for a root is a small house like structure over the number with a lightly smaller number raised to left to indicate the power of the root.

$$\sqrt[3]{8} = 2$$

read "the cubed root (or the 3rd root) of 8 equals 2 is the inverse of $(2)^3 = 8$.

Again, we can write a general definition using variables:

If x and b are any real numbers

Then $\sqrt[b]{(x^b)} = x$

Inverse Operation for Exponents (Roots)

1.3.3 Absolute Value

We might note that the definition does not address the sign of the result as it is embedded in the variable. Sign is, however, quite important when considering roots. We know that when a negative number is multiplied by itself it produces a positive result. Since the identical positive number multiplied by itself will produce the identical positive result, it is easy to see that the root operation actually has two possible answers for every square root – a positive one and a negative one of the identical magnitude. This is true when whenever we take a root of any even power.

Oftentimes we are only interested in the positive solution for a root operation. When we are using variables to write general expressions, we need a way to show that we are only interested in the positive part of a number as the sign is hidden in the variable and as yet unknown. For this purpose another operation, the Absolute Value, has been defined.

The Absolute Value is an operation that always returns the positive value of a number, or its magnitude. Its symbol is $|\ |$. We cannot write a general definition for Absolute Value since, as previously discussed, the sign is included in the variable. The examples below will illustrate.

$$|-9| = 9,\ |+9| = 9$$

$$|-3.5| = 3.5,\ |+3.5| = 3.5$$

1.3.4 Rules of Exponents

The most powerful feature of exponential format is not, as suggested previously, in the memorization of number facts but reveals itself instead when we combine multiplication and division problems with exponential format. For instance, we might note that, by definition,

$$3^2(3^4) = (3 \cdot 3)(3 \cdot 3 \cdot 3 \cdot 3)$$

If we then count all the threes being multiplied we will note that we have 6. We may also note that the 2 and the 4 in the exponents on the left add to the 6, the exponent on the right, and that this makes complete sense because each of the exponents is just keeping track of how many 3s there are. It is clear to see that no matter what the base or what the exponents, if the bases are the same and we are multiplying, we need only add the exponents to keep track of the result.

Stated as a general rule:

> If x, a, and b and are any real numbers
>
> Then $x^a(x^b) = x^{(a+b)}$

We can use the definition of exponents in a similar manner for division problems,

$$\frac{6^7}{6^4} = \frac{6(6)(6)(6)(6)(6)(6)}{(6)(6)(6)(6)} = \frac{6}{6}\left(\frac{6}{6}\right)\left(\frac{6}{6}\right)\left(\frac{6}{6}\right)\left(\frac{6}{1}\right)\left(\frac{6}{1}\right)\left(\frac{6}{1}\right) = 1(1)(1)(1)(6)(6)(6) = 6^3$$

And write a similar general rule:

> If x, a, and b and are any real numbers
>
> Then $\frac{x^b}{x^a} = x^{(b-a)}$

These general rules can help us to significantly simplify complicated problems yielding both faster and more accurate results.

Example

The mass of the earth is approximately $6*10^{24}$ kg. The approximate mass of one atom is $1.7*10^{-27}$ kg. About how many atoms are there in the earth?

To find the number of atoms in the earth we must take the mass of the earth and divide by the mass of an atom as shown below.

$$\frac{6*10^{24}\,kg}{1.7*10^{-27}\,kg/atom}$$

If we notice that both the exponential forms have the same base, we can quickly rewrite the problem:

$$\frac{6}{1.7}*\left(10^{[24-(-27)]}\right)\cdot atoms = \frac{6}{1.7}*10^{51}\,atoms$$

Ten raised to the 51st power, a 10 with 50 trailing zeroes, is a number that cannot be described easily with the English language, yet with the power of general rules for exponents we can express it mathematically and use it with ease in calculations.

Exponents of Exponents

Keeping with the same analysis we can expand our applications. What if we need to multiply 2^2 by itself four times? We can express this as

$$2^2(2^2)(2^2)(2^2)=(2^2)^4.$$

When we have such an expression, we can use the definition of exponents twice to resolve the total number of times we are multiplying. For instance,

$$(3^2)^3=3^2(3^2)(3^2)=(3\cdot 3)(3\cdot 3)(3\cdot 3)=3^6=3^{(2\cdot 3)}.$$

We can use this analysis to readily write another rule for exponents:

If x, a, and b equal any real numbers

Then $\left(x^{a}\right)^{b} = x^{(ab)}$

Interesting things that stem from the rules for exponents

What do you get if you multiply something by itself zero times? You obviously can't get zero since neither of the factors is zero. You can't get the number itself back since that would be the number multiplied by itself 1 time x^1. A clue lies in the rule for dividing exponents. We know that

$$\frac{x^{a}}{x^{a}} = 1$$

Since anything divided by itself is one. If we apply the division rule, however, we find

$$\frac{x^{a}}{x^{a}} = x^{(a-a)} = x^{0} = 1.$$

What if we would like to get rid of those silly little root houses? Is there a way to express the inverse of an exponent as an exponent? Close inspection shows this to be possible. We are looking for an exponential form for the classical form $\sqrt[b]{x}$. To find it we remember that

$$\sqrt[b]{\left(x^{b}\right)} = x.$$

Substituting an exponential form for the root yields:

$$\sqrt[b]{\left(x^{b}\right)} = \left(\left(x^{b}\right)\right)^{a} \quad \text{(we are looking for a).}$$

Substituting the first expression into the second we find

$$\sqrt[b]{\left(x^{b}\right)} = \left[\left(x^{b}\right)\right]^{a} = x = x^{1}. \qquad \text{Knowing that } \left[\left(x^{b}\right)\right]^{a} = x^{(ab)},$$

We are looking for an expression for a that will make ab = 1. It is obvious that

$$a = \frac{1}{b} \quad \text{since} \quad \left(\frac{1}{b}\right)b = 1.$$

We can now write a general statement that

$$\sqrt[b]{x} = x^{\frac{1}{b}}.$$

As you will see, using exponential form for roots in mathematical calculations not only makes them easier to read and write, it streamlines many exponent and multiplication operations where roots are involved.

In summary, we can use variables to write rules that make exponents easier to work with. A summary of these rules can be found below.

Rules for Exponents

If x, a, and b are any real numbers

Then $x^a(x^b) = x^{(a+b)}$

$$\frac{x^b}{x^a} = x^{(b-a)}$$

$$(x^a)^b = x^{(ab)}$$

$$x^1 = x$$

$$x^0 = 1 \quad \text{and,}$$

$$\sqrt[b]{x} = x^{\frac{1}{b}}.$$

1.3.5 Puzzling It All Out

Learning to play a musical instrument, build a canoe, or any other worthwhile endeavor requires active involvement before the information provided can be fully integrated. The study of mathematics is no different.

Work each of these puzzles on your own until you are sure you have the correct answer. Then check the answer in the back of the book. If you are having trouble agreeing with the book's answer you can: check the solutions manual if one is available in your area or ask an online program to show you a solution, ask a friend or a tutor, or ask your teacher for advice. Keep in mind that there may be more than one way to solve each problem. Always look for the easiest solution.

It is preferable that you understand how to get the correct solution for every puzzle on this page before moving to the next section.

Evaluate by both definition and rules – the first is done as an example:

1. $y^4 \cdot y^2 = (y \cdot y \cdot y \cdot y)(y \cdot y) = y^{(4+2)} = y^6$

2. $w^3 \cdot w^5 =$

3. $\dfrac{a^7}{a^4} =$

4. $(-2)^3 \cdot (-2) =$

Write:

5. Three different ways of mathematically saying 1 using three different operations.

6. A general statement stating that squaring a negative number results in a positive number.

Food for Thought

How could we write a statement that any raising a negative number to any even power results in a positive number?

1.4 Combined Problems

We have found that addition and multiplication can be done in any order and have used this property (the Commutative Property) to help us solve problems. We must now consider what happens when addition and subtraction occur together in a single problem or statement. Great care must be taken when approaching these types of problems. Understanding combined problems is easier when we recall that both multiplication and exponential format are shortcuts that have been developed from addition. Each of them has their own unique symbolic form. When we read numbers (or letters) and symbols in a mathematical statement, we must keep this in mind. For instance, we might look at the expression below

$$3+6(7)$$

and be tempted to work from left to right, adding the 3 and 6, getting 9, and then multiplying by 7 resulting in an answer of 63.

However, if we first translate the expression, converting it all into addition,

$$3+6+6+6+6+6+6+6,$$

we will see immediately that the first approach is in error for the statement is obviously showing seven sixes, not seven nines and when we do the addition we realize the answer to be 45, not 63.

Had we commuted the addends to do the multiplication first we would have evaluated it as

$$6(7)+3=42+3=45, \quad \text{the true answer.}$$

A similar situation arises when exponents are combined with multiplication or addition. I have seen students want to evaluate the expression

$$(3)4^2$$

from left to right coming up with 12^2 or 144.

When we translate the expression back to multiplication,

$$3(4)(4)$$

we see the error in this approach as the real answer is clearly 64.

Commuting to bring the exponential form to the left again provides an approach with the correct answer.

$$4^2(3)=16(3)=64.$$

1.4.1 Order of Operations

Since higher level functions were developed to save time, translating expressions into like operations is counterproductive. The use of the commutative property to perform higher order operations first as shown above is a far more efficient solution. However, long or complicated expressions may be difficult to analysis in this manner. For instance what can we commute in the expression

$$3+\frac{4^2}{2^2}-(4)(-3)^3\,?$$

Not surprisingly, we have developed a method that efficiently answers this question. Each higher order operation is converted to its fundamental state not through translation but through performing the operation, in place, before evaluating the expression. After each successive higher order operation has been evaluated, the expression can be evaluated from left to right (or in any order) like any addition (subtraction) problem.

Obviously exponents (and roots) must be evaluated prior to multiplication since the operation includes multiplication. Likewise, multiplication and division, which are repetitive addition and subtraction, must be done before addition and subtraction.

The idea gives rise to an order of operations that, when followed, will always yield the correct result. For the operations we have studied thus far, this order is as follows (first listed, first performed).

Order of Operations

Exponents and Roots
Multiplication and Division
Addition and Subtraction

Under the order of operations, the expression above would be successively evaluated as shown below.

$$3+\frac{4^2}{2^2}-(4)(-3)^3=3+\frac{16}{4}-4(-27)=3+4+72=79\,.$$

1.4.2 Associative Property

Sometimes, we may specifically want to perform a lower level operation prior to a higher level one. To do this, we must have a way to express it symbolically in our expressions. We call this "operation" Association and its symbol is ( ).

Example

Mrs. Jones brings 4 children to the art class on Tuesdays. Mr. Todd brings 2, and Maynard brings 1. If the instructor for the art class charges $5 per child per class, how much money will she have collected after 6 classes?

There are two ways to approach this problem. One is to multiply each group of children by 6 to find how many classes they have attended and then add the total for all the children. It looks like this in a mathematical expression:

$$4(6)(\$5) + 2(6)(\$5) + 1(6)(\$5) = \$120 + \$60 + \$30 = \$210$$

If we do so we will notice that we have multiplied by both 6 and 5 three times. As mathematicians, we immediately start looking for an easier way!

We are saved by Association. We can specify to add the number of children that come to one class first using parenthesis. Then, we need multiply the total by 6 and $5 once.

$$(4+2+1)(6)(\$5) = (7)(6)(\$5) = \$210$$

Ah...... much better.

Association has come to be called the Associative Property, although it is really more of an operation than a property. To make it work, the order of operations must be amended so that anything being associated is performed first. A acronym, PEMDAS (parenthesis, exponents, multiplication, division, addition, subtraction) has been developed that can help you to remember the order. If you forget PEMDAS, just remember to do higher order operations prior to lower order ones with any associated quantities done before all.

Implied Association

In some cases, parentheses are not needed because the association is already implied. Division problems (or "fractional forms") often include this type of implied association.

Example

Jack collected money for his band's costumes in three separate fundraisers netting $400, $510, and $390 respectively. He must divide the money up between 5 band members. How much money does each member get?

The approach is obvious. You total up the money received and then divide by 5. The proper way to write it as a math expression is:

$$\frac{(\$400+\$510+\$390)}{5}=\frac{\$1300}{5}=\$260$$

However, since the division sign makes a natural break, we often omit the parenthesis.

$$\frac{\$400+\$510+\$390}{5}=\frac{\$1300}{5}=\$260$$

Remember, mathematicians are by their nature lazy (why write an extra set of parenthesis if you don't have to? And anyway, it's easier to read. Be wary of implied parenthesis. Trying to perform division prior to evaluating a dividend or divisor can lead to highly erroneous results.

1.4.3 The Distributive Property

Order of Operations can help us evaluate expressions containing only constants, but it is of little help when our expressions contain variables since we cannot evaluate the higher order expressions. While

$$3+2\left(3^2\right)$$

can be evaluated readily as,

$$3+2\left(3^2\right)=3+2(9)=3+18=21,$$

$$a+2(b)^2$$

cannot be further evaluated at all.

There are, however, expressions that just beg for a method to be developed because of their repetitive patterns. For example, in the expression that we developed for the children in the art class, we had a 6 in every term,

$$4(6)(5)+2(6)(5)+1(6)(5).$$

Since we were creating the expression, we chose to write it in a different manner that associated the numbers that were not repetitive,

$$(4+2+1)(6)(5),$$

with both expressions arriving at the same answer, 210. It is obvious, however, that there is a pattern here that could have been employed directly without rewriting. When these same expressions are written with variables, a new property is revealed.

For a, b, c, and f equal to any real number

$$ab+ac+ad=a(b+c+d)$$

Distributive Property

Class Activity

All class members stand in parallel lines forming 2 or 3 lines depending on the number of students. Leave a gap in each line with the same number of students on each side of the gap in each line (see diagram below).

One student from each side of the break writes an expression on the board that shows the total number of students on their side of the break, for example

$$2(3) = 6.$$

The student on the other side of the break will have a similar expression, possibly

$$2(2) = 4.$$

A different student should then write an expression for the total number of students by looking down the lines. In the case of the configuration above, it will look like this:

$$2(3 + 2),$$

showing 2 lines with a group of 3 and a group of 2 students in each of them.

If done correctly, the two expressions will obviously add to the same amount (the number of students standing). In this case we have:

$$6 + 4 = 2(5) = 10$$

The students should then write a general statement for the property that they have just observed using variables. We call this property the Distributive Property.

[Note: If you are using this book at home and do not have enough friends to make the lines, you can use ceramic figurines, scrabble pieces, or (of course) toothpicks placed where the students would stand. Then just write each of the expressions yourselves.]

1.4.4 Addition and Subtraction of "Fractions" or Distributive Property for Rational Forms

You have already seen how the commutative property for rational forms (division problems) leads naturally to the rules and methods developed for fractions. In a similar manner, the Distributive Property for Division leads directly to the rules taught for fractions. To make the correlation obvious, we will use rational forms that result in whole numbers in this discussion. Many arithmetic texts refer to these as "improper fractions".

Example

Evaluate the expression $\frac{4}{2}+\frac{6}{2}$.

There are two obvious methods of approach based on prior discussions.

Since there are only constants, we can evaluate the expression using the Order of Operations. Performing the division first we can easily compute:

$$\frac{4}{2}+\frac{6}{2}=2+3=5.$$

We could, however, use the Distributive Property to factor out the common $\frac{1}{2}$ (or division by 2). We can then just add the sum that we are left with and divide by the common 2 when we are done.

$$\frac{1}{2}(4+6)=\frac{1}{2}(10)=\frac{10}{2}=5.$$

Using the Distributive Property makes little sense for the problem above since it can readily be solved through the Order of Operations, but it provides us with a tool to solve problems that are not so easy such as:

$$\frac{7}{3}+\frac{8}{3}=.$$

Here, we are faced again with infinitely repeating decimals. Performing the division first will result in an approximate answer at best. In this case, a quick application of the Distributive Property will produce the desired result:

$$\frac{7}{3}+\frac{8}{3}=\frac{1}{3}(7+8)=\frac{15}{3}=5.$$

When learning to manipulate these forms in Elementary School, many of us learned this as a rule for adding fractions. We were told that we could just add the numerators (dividends) if the denominators (divisors) were the same. Although this method is as effective as (an in fact the same thing as) using the Distributive Property, we will not refer to it in this book with this terminology (other than in this Section).

In those same Elementary School classes, we were taught that if the denominators were not the same then we must find a "common denominator". We do this by changing the form of each fraction (or rational form) so that each one has the same divisor. We can then, again, use the Distributive Property to factor out the common divisor so that we can just add the dividends (numerators).

Example

Evaluate $\frac{6}{2}+\frac{9}{3}=$

First, we should notice that direct application of the Order of Operations will solve the problem directly:

$$\frac{6}{2}+\frac{9}{3}=3+3=6.$$

We could, however, obtain like divisors by multiplying each rational form by an appropriate form of 1 as shown below.

$$\frac{6}{2}+\frac{9}{3}=\frac{6}{2}\left(\frac{3}{3}\right)+\frac{9}{3}\left(\frac{2}{2}\right)=\frac{18}{6}+\frac{18}{6}.$$

The common 6 can then be factored out using the Distributive Property and the answer quickly obtained:

$$\frac{18}{6}+\frac{18}{6}=\frac{1}{6}(18+18)=\frac{36}{6}=6.$$

We do not, of course, have to physically factor out the common divisor. We can easily just add the dividends (numerators) as we were taught (or since we know that the divisors are the same). The steps are added here to clarify that the rules that we have developed for fractions are not separate entities that apply only to a special form, but are specializations of one of the basic properties of mathematics, the Distributive Property, as applied to rational expressions.

The Distributive Property is an especially useful tool when dealing with variables. We can quickly use this property to expand or factor an expression as needed. These situations arise often when working with algebraic expressions.

Example

Evaluate $$\frac{a(2+b)+b(2+a)}{(a+b+ab)}$$

We can use the Distributive Property, $ab+ac+ad=a(b+c+d)$, on the dividend to expand each of the terms as shown below:

$$\frac{a(2+b)+b(2+a)}{(a+b+ab)}=\frac{2a+ab+2b+ab}{(a+b+ab)}$$

We can then use it again (after commuting and combining) to obtain:

$$\frac{2a+ab+2b+ab}{(a+b+ab)}=\frac{2a+2b+2ab}{(a+b+ab)}=\frac{2(a+b+ab)}{(a+b+ab)},$$

$$=2(1)=2$$

greatly simplifying the expression.

1.4.5 Expanded Distributive Property

Class Activity

If people are used as markers, this activity will require 45 students. Ceramic figurines, scrabble pieces, or toothpicks may again be substituted. Place the markers in four groups aligned and broken as shown below:

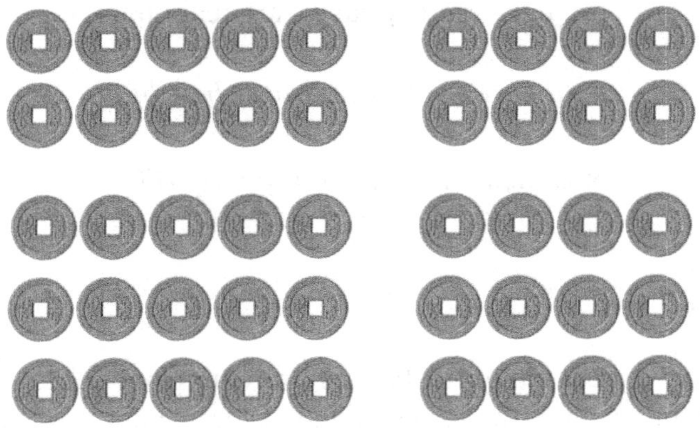

Step 1: As before, write an expression for each of the four groups (ie. 5(2) = 10) and find the sum of all of the groups combined.

Step 2: Look at the markers again and write an expression that calculates the number of markers by multiplying the numbers in the columns by the number of rows. Indicate the group breaks by using an associated sum for the number of markers on each side. For example, the top row would be written as (5 + 4), not 9.

Step 3: Since the two expressions both equal the total number of markers, they are equal to each other. Write an expression showing this.

The result will look something like this:

Step 1: $2(4)+2(5)+3(4)+3(5)=8+10+12+15=45$

Step 2: $(2+3)(4+5)=5(9)=45$

Step 3: $(2+3)(4+5)=4(3)+5(3)+2(4)+2(5)$

If we are careful, we can write a general expression showing the property we have just observed as shown below.

For a, b, c, and f equal to any real numbers

$$(a+b)(c+f)=ac+af+bc+bf$$

Expanded Distributive Property (FOIL)

It can be observed that expansion of the factors (moving from left to right), can be achieved by multiplying each of the terms in one of the factors by each of the terms in the other. To assist with remembering all terms, the acronym FOIL was developed indicating to multiply the **F**irst terms in each factor (a & c), the **O**uter terms between the 2 factors (a & f), the **I**nner terms between the two factors (b & c), and the **L**ast terms in each factor (b & f). Arrows drawn from each term in one factor to the other terms in the other are often used also or instead as shown below.

$$(a+b)(c+f)=ac+af+bc+bf$$

1.4.6 Perfect Squares and the Difference of Squares

Two special cases of the Expanded Distributive Property are of special interest as they occur often and recognizing them may facilitate problem solving. Perfect squares occur when we square a binomial (a combination of two terms). Using the Expanded Distributive Property from left to right we can easily find the product in general, expanded form:

$$(a+b)^2=(a+b)(a+b)=a^2+ab+ab+b^2=a^2+2ab+b^2,$$

or

$$(a+b)^2 = a^2 + 2ab + b^2$$

Perfect Square Expansion

The Difference of Squares is the result of multiplying a sum of two numbers by a difference of the same two numbers as shown below.

$$(a+b)(a-b) = a^2 - ab + ab + b^2 = a^2 - b^2,$$

or

$$(a+b)(a-b) = a^2 - b^2$$

Difference of Squares

Example

Evaluate $\dfrac{x^2 - y^2}{x + y}$.

If we recognize $x^2 - y^2$ as the difference of two squares, we can immediately transform it into its factored form, $(x+y)(x-y)$. Our expression then becomes

$$\frac{x^2 - y^2}{x + y} = \frac{(x+y)(x-y)}{x+y}.$$

$$= \frac{x+y}{x+y}(x-y) = 1(x-y) = x - y$$

1.4.7 Puzzling It All Out

Learning to play a musical instrument, build a canoe, or any other worthwhile endeavor requires active involvement before the information provided can be fully integrated. The study of mathematics is no different.

Work each of these puzzles on your own until you are sure you have the correct answer. Then check the answer in the back of the book. If you are having trouble agreeing with the book's answer you can: check the solutions manual if one is available in your area or ask an online program to show you a solution, ask a friend or a tutor, or ask your teacher for advice. Keep in mind that there may be more than one way to solve each problem. Always look for the easiest solution.

It is preferable that you understand how to get the correct solution for every puzzle on this page before moving to the next section.

Evaluate the combined expressions. Leave all answers in lowest whole number form:

1. $\frac{1}{3}+\frac{1}{2}=$

2. $4w^3+3w^3-4w=$

3. $\frac{a^2}{x}+\frac{a^2}{y}-\frac{2a^2}{z}=$

4. $\sqrt{f}\left(f^3\right)=$

5. $\frac{x^2y+y^2x}{y^3x}=$

6. $\frac{4x^2-12xy+9y^2}{(2x-3y)}=$

Write:

7. A general method showing the addition of rational forms. (Hint, do it through with constants and then substitute variables.)

1.5 Chapter 1 - Review

We were told in the introduction that we can use variables in mathematics to make General Statements, explore Relationships, and solve for unknowns.

In this Chapter we have looked at how using variables to write general statements can help us understand the rules of arithmetic. We identified several properties that can help us evaluate both arithmetic and algebraic expressions quickly and easily.

1.5.1 General Rules for Addition, Subtraction, Multiplication, and Division

For x, y, a, and b equal to any real numbers:

We have identified that

Addition is a shortcut for counting;

Subtraction is the inverse of Addition,

$$x + y - y = x;$$

Multiplication is a shortcut for repetitive Addition,

$$x + x + x = 3x;$$

and Division as the inverse of Multiplication,

$$\frac{xy}{y} = x.$$

We have observed the following properties for these operations:

Identity Properties of Addition and Multiplication

$x + 0 = x$ Addition

($x - 0 = x$ Subtraction)

$x(1) = x$ Multiplication

($\frac{x}{1} = x$ or $\frac{x}{x} = 1$ Division),

Zero Product Property for Multiplication

x(0) = x

or

If ax = 0 Then a = 0 or x = 0

Commutative Properties of Addition and Multiplication

x + y = y + x or x – y = - y + x

and

$$\frac{a}{b}\cdot\frac{x}{y}=\frac{x}{b}\cdot\frac{a}{y}=\frac{a}{y}\cdot\frac{x}{b}$$

Inversion of Divisor Property for Division by Rational Forms

$$\frac{\left(\frac{a}{b}\right)}{\left(\frac{x}{y}\right)}=\left(\frac{a}{b}\right)\left(\frac{y}{x}\right).$$

1.5.2 Exponents

We have identified exponential form as a shortcut for repetitive multiplication, $x \cdot x \cdot x = x^3$ and its inverse, the root ($\sqrt[a]{x^a} = x$). We also observed the following rules to be true for exponential forms:

Rules for Exponents

$$x^a x^b = x^{(a+b)}$$

$$\frac{x^a}{x^b} = x^{(a-b)} \qquad x^0 = 1$$

$$\left(x^a\right)^b = x^{ab} \qquad \sqrt[a]{x} = x^{\frac{1}{a}}$$

1.5.3 Combined Operations

We observed that the mathematical symbolism for higher order operations such as exponents or multiplication can cause problems when trying to evaluate mixed expressions. We observed that performing higher order operations prior to lower order operations resolves this problem when it is possible to do. We also observed that, especially when using variables, that this practice is not always preferable or possible. We looked for properties that would help us with this and found two:

The Associative Property

Not a true property but an operation itself, we can indicate that any part of an expression is to be evaluated first by separating it with parenthesis.

The Distributive Property

$$ab + ax + ay = a(b + x + y)$$

We showed that we could expand the distributive property to handle two, or more, factors of sums and that some of these are seen often enough to make them worth recognizing as shown below:

Expanded Distributive Property (FOIL)

$$(a + b)(c + f) = ac + af + bc + bf$$

Perfect Square Expansion

$$(a + b)^2 = a^2 + 2ab + b^2$$

Difference of Squares Expansion

$$(a + b)(a - b) = a^2 - b^2$$

1.5.4 More Practice Puzzles

Evaluate the combined expressions. Leave all answers in lowest whole number form. Try to find the easiest solution possible (answers in back of book):

1. $\frac{x}{x-2}-\frac{2(x+2)}{(x^2-4)}=$

2. $([(x-y)(x+y)+y^2])^{\frac{1}{2}}$

3. $\frac{y}{3}+\frac{3}{y}=$

4. $\sqrt{3}\left(\frac{3^2+6^2}{9\sqrt{12}}\right)=$

5. $\frac{(x^2y)^3+y^3x}{y^3x}=$

6. $\frac{\frac{4x^2-2xy-(-9x^2)}{5x}}{\frac{13x-2y}{25}}=$

Chapter 2 – Formulas

Formulas are another powerful usage of variables in mathematics. A formula is a rule that we have developed for a given situation that occurs often. Instead of solving the problem independently every time with numbers, we can solve it once with variables and develop a formula. Once a formula is developed, the numbers for a given situation can be substituted back into the formula and an answer obtained directly. Many formulas have been developed in every area of life. We will look at only a few of them here as examples.

2.1 Geometric Formulas

Most of us have developed formulas for the perimeter or area of a rectangle or looked at other geometric shapes so this is a good place to start. Developing a formula for a geometric situation is usually best approached by drawing a picture. Important parts of the picture can be labeled with variables and the variables used to develop the equation.

2.1.1 Rectangles

A rectangle is a four sided planar object that has two sets of identical length opposing sides and right angles at each corner as shown below. (The boxes in the corners indicate a right, or 90° angle.)

Rectangle

If we want to know the distance around the rectangle, it is obvious that we must measure each of the sides and add the distances along each side together. If we remember that each of the opposing sides are the same, we can easily substitute variables into this procedure to obtain a more general procedure. The labeled rectangle and the resulting formula are shown below.

For P = Perimeter (distance around) of a rectangle,
L = Length (as shown), and W = Width (as shown)

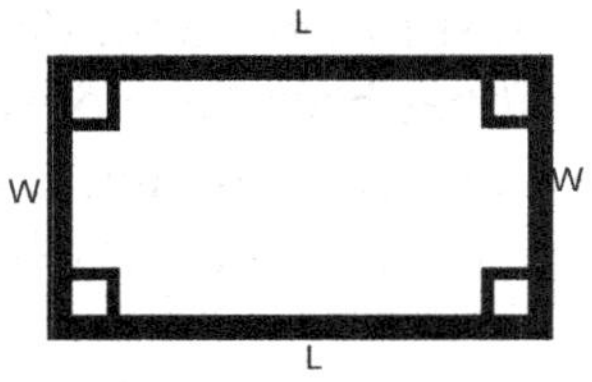

Then $P = L + W + L + W$ as discussed.

We might also observe that the expression contains repetitive variables. We can use the commutative property to rearrange and the distributive property to add like terms to come up with a simple, easy to use expression for the perimeter of any rectangle.

$$P = L + L + W + W$$

$$P = (1+1)\,L + (1+1)W$$

$$P = 2L + 2W$$

$$P = 2(L + W)$$

Once we have this formula we can easily obtain the distance around any rectangle by measuring the length and the width and substituting into the formula.

Example

Find the distance around a rectangle with a length of 6’ and a width of 2’.

We can assign numbers directly to the length and width variables because we are given these.

$$L = 6' \quad \text{and} \quad W = 2'$$

Keeping in mind that the distance around is the perimeter (P), we can calculate this value directly by substituting the other two into the formula.

$$P = 2L + 2W = 2(6') + 2(2') = 12' + 4' = 16'$$

We can perform a similar procedure to find a formula to represent the area (the surface the covered) of any rectangle. We must first have (or develop) a mathematical understanding of what area is. A unit has been developed to describe, and quantify, surface area and is called a square unit. A square unit is defined as an area that is 1 unit long by 1 unit wide as shown below. Units can be any linear measurement (inches, feet, miles, etc).

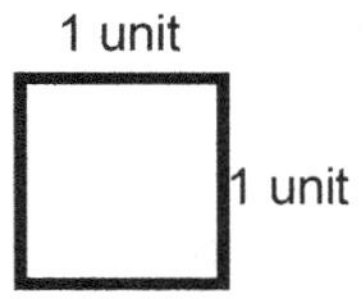

One Square Unit

The total area of any object is then described as the number of square units that the object occupies. The number of square units in any rectangle can be quickly observed by inserting the square unit measure into the rectangle of interest. For the rectangle in our earlier example, it would look like the drawing below with each unit equal to 1 foot.

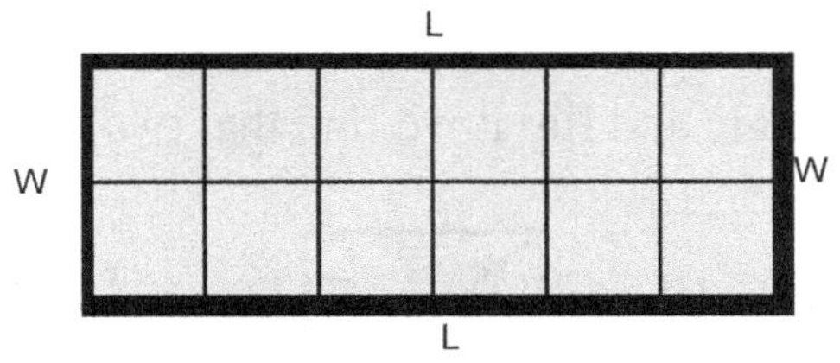

We can count the square units and see that there are a total of 12, in this case 12 square feet. We could note that we could also arrive at twelve square feet if we multiplied the length, 6', by the width, 2'. The diagram is, in fact, exactly like diagrams many of us have seen developed to show what multiplication is. We obviously have 2 rows of 6 items each. In this case, the items are square units and we come to the unmistakable conclusion that we can find the number of them that will occupy any rectangle by simply multiplying the length of the rectangle by its width.

If we use the variables defined above (L = Length and W = Width) and define A = the area of a rectangle then we can write a simple formula for the area of any rectangle as:

$$A = LW.$$

2.1.2 Triangles

A triangle is a planar geometric shape with 3 sides. To find a formula for the distance around the triangle we can use a similar procedure to that for rectangles.

For a triangle with sides a, b, and c as shown below:

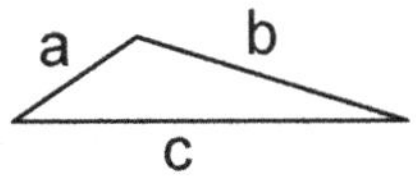

The distance around the triangle (the perimeter of the triangle) = P_T

$$P_T = a + b + c$$

In the case of a triangle we cannot simplify this expression further.

A method to determine a formula for the area of a triangle is not as readily apparent. At first glance, it would appear impossible to easily calculate a number of square units in such a random shape. However, if we start with a somewhat less random shape, we may be able to extrapolate a formula that can be generalized to all. A somewhat obvious place to start might be a right triangle, a triangle with one right angle as shown below.

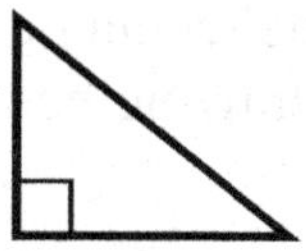

If we duplicate the right triangle and flip it we find that two right triangles make a rectangle:

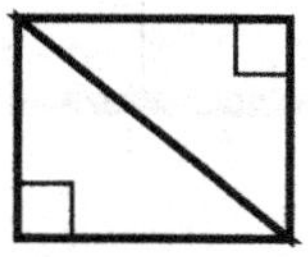

The area of a right triangle is then one half the area of a rectangle if we use the lengths of the two sides adjacent to the right angle as the length and the width in the area of a rectangle formula. To be sure that we use the proper sides, we have developed specific names for them. One side is called the base, often denoted by b. The other side, the one perpendicular to the base is called the height and is usually represented by h. Using these variables and A_T to represent the area of the right triangle, we can state that

$$A_T = \frac{1}{2}LW = \frac{1}{2}bh$$

Where A_T = Area of the triangle, b = base, and h = height (as defined above).

As it turns out, any triangle can be turned into two right triangles by adding a line perpendicular to one side. The area of any triangle can then be found using a similar analogy.

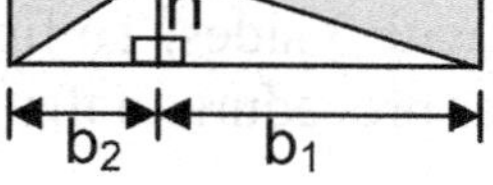

The area for each of the triangles can be found using the formula derived for a right triangle and then added together to give

$$A_T = \frac{1}{2}b_1 h + \frac{1}{2}b_2 h$$

We can then use the distributive property to gather the common $\frac{1}{2}$ and h to get

$$A_T = \frac{1}{2}h(b_1 + b_2).$$

If we notice that (b1 + b2) is equal to the length of the side that is perpendicular to the height, we will realize that the formula developed earlier for a right triangle can be generalized to any triangle by defining the

height as a line drawn from any vertex (corner) to the opposite side of the triangle forming a right angle
and the
base as the length of the side that the height intersects as shown below.

For any triangle with height and base as labeled below,

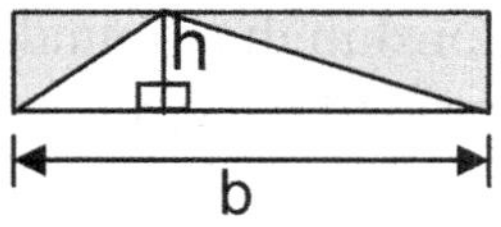

The area of the triangle, A_T, can be found with the formula

$$A_T = \frac{1}{2}bh$$

Example

Find the area of the triangle shown below:

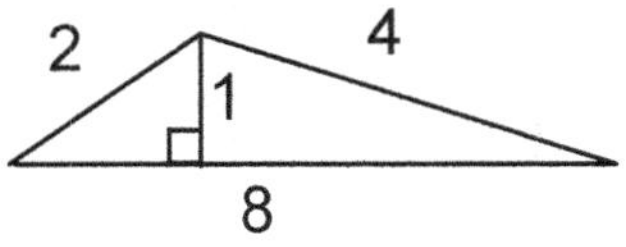

Although all of the lengths are given, we only need the length of the base and the height. The height, by definition, goes from a vertex and makes a right angle with the opposite side. We can easily identify the length of the line that meets this requirement as 1 unit. The base is then the entire length of the side it intersects, in our case 8 units. The area of the triangle can then be found using the developed formula as:

$$A_T = \frac{1}{2}bh = \frac{1}{2}(8)(1) = \frac{8}{2} = 4$$

2.1.3 Circles

Many years ago the Greeks developed formulas for the distance around (called the circumference) and the area of a circle based on an interesting relationship that they found between the circumference and the distance across at the center (called the diameter). Observation and careful measurement of many circles revealed that the ratio of the circumference to the length of the diameter was always the same. Like many constants found in nature, however, the resulting number is irrational (goes on forever). Since we cannot write out an irrational number in mathematical statements, the Greeks assigned a symbol, the Greek letter π, to represent this very important constant.

In mathematical language, the relationship they found was that for a circle labeled as below and the distance completely around the circle equal to the circumference (C),

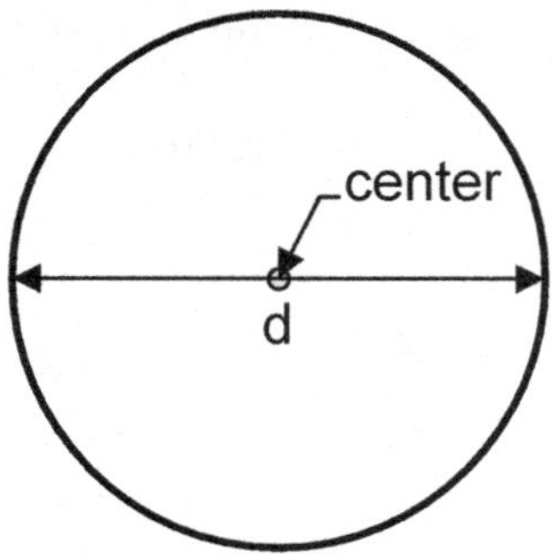

$$\frac{C}{d} = \pi.$$

From this relationship they developed the following two formulas for the circumference and the area respectively.

$$C = \pi d \qquad \text{and} \qquad A = \pi r^2$$

Where C = the circumference, A = the area,

d = the diameter, and $r = \frac{d}{2}$ or $\frac{1}{2}$ the diameter.

We will see how the first formula was derived in Unit 2 (although you might have already guessed). The second requires more geometric analysis than we will cover in this text but is an interesting topic to research elsewhere.

2.1.4 Puzzling it All Out

Learning to play a musical instrument, build a canoe, or any other worthwhile endeavor requires active involvement before the information provided can be fully integrated. The study of mathematics is no different.

Work each of these puzzles on your own until you are sure you have the correct answer. Then check the answer in the back of the book. If you are having trouble agreeing with the book's answer you can: check the solutions manual if one is available in your area or ask an online program to show you a solution, ask a friend or a tutor, or ask your teacher for advice. Keep in mind that there may be more than one way to solve each problem. Always look for the easiest solution.

It is preferable that you understand how to get the correct solution for every puzzle on this page before moving to the next section.

1. Find the perimeter of the figure below
   [Hint: the Pythagorean Theorem - $c = \sqrt{a^2 + b^2}$ (where c equals the longest side of a right triangle and a and b are the other two sides) can be used to calculate the unknown length.]

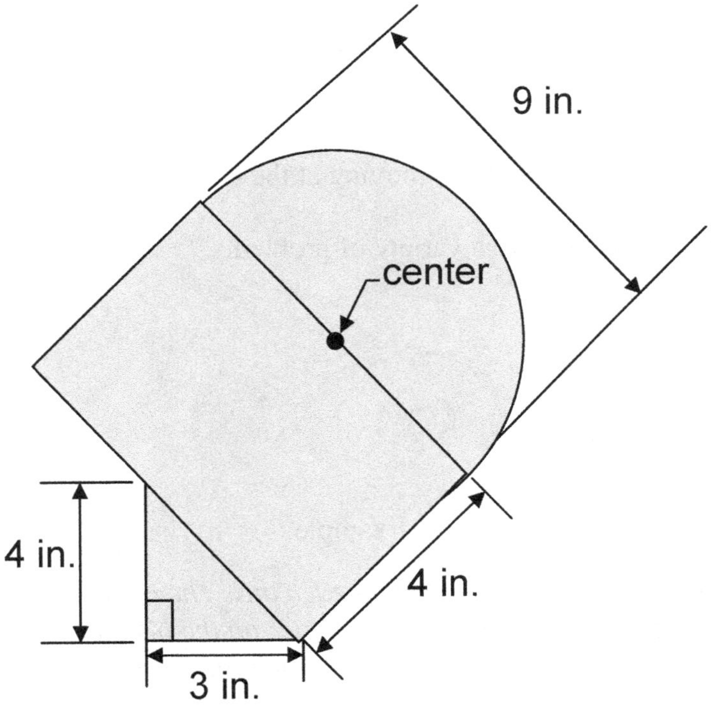

2. What is the area of the figure in puzzle 1?

3. Develop a formula for the area and perimeter of a square (a planar geometric figure with 4 sides of equal length). How does this formula differ from that for those developed for a rectangle? Why?

2.2 Scientific Formulas

Just as geometric formulas help us solve geometric problems quickly and easily, scientific formulas help us to solve problems in the world of science. In addition, formulas are developed in science to communicate things we have learned about our world.

As with geometric formulas, labeled pictures are often helpful when developing scientific formulas. These formulas are also built from data collected during scientific observation.

2.2.1 Distance, Rate, and Time

One area of Physics, Dynamics, studies physical bodies in motion. The formulas developed through these studies help us understand things like the movements of planets and stars, what happens when gases collide, or why the fender on your new automobile quickly changed significantly in size and shape. The study of Dynamics begins with the simple observation that an object moving at a particular rate (speed or velocity) for a particular amount of time will cover a distance than can be predicted by the formula

$$d = rt$$

Where

d = the distance covered, r = the rate (or speed) the object is traveling, and t = the time the object is moving at the given rate.

This formula can be used to solve a variety of problems.

Example

Sally rides her bicycle along a smooth path each day. The path is 25 miles long but she seldom rides it completely. She has a speedometer on the bike and has found that she is always able to keep the same steady 10 mph along the path. Last week she rode for 2 hours on Monday, an hour and a half on Tuesday and Thursday, and for two and a half hours on the weekend. She is keeping a log of her physical activities and needs to know how many miles she bicycled in total last week.

To solve this problem we must use the distance, rate, time formula to find the distance traveled on each day. The total distance is then the sum of the distances.

For all days, r = 10 mph.

On Monday, t = 2 hrs

So $$d_{Monday} = 10\frac{miles}{hour}(2hours) = 20miles$$

On both Tuesday and Thursday, t = 1.5 hrs.
We can use the formula once and double the result.

So $$d_{TuesandThurs} = 2d_{Tues} = 2\left(10\frac{miles}{hour}\right)(1.5hours) = 30miles$$

On the weekend, t = 2.5 hrs

And $$d_{Weekend} = 10\frac{miles}{hour}(2.5hours) = 25miles$$

The total miles is the sum of the three

$$d_{Total} = d_{Monday} + d_{TuesandThurs} + d_{Weekend} = 20 + 30 + 25 = 75 \text{ miles.}$$

2.2.2 Temperature Scales

We have developed several different scales to measure temperature. The most common of these are the Fahrenheit Scale, developed for the English unit system and the Celsius or Centigrade Scale developed for the metric system. We are often given data in units from one scale when we need it for the other. A formula that can convert between the two quickly and easily is highly desirable.

To develop such a formula we need to find a relationship between the two scales that can be expressed mathematically. To help with this, we will look at a picture of the two scales side by side. Here, we have used two known temperatures to align the scales; the boiling point of water (100°C, 212°F) and the freezing point of water (0°C, 32°F).

Celsius	Fahrenheit
100	212
90	194
80	176
70	158
60	140
50	122
40	104

30	86
20	68
10	50
0	32

Although a more analytical analysis is possible, we will try a simple logical solution. With the scales aligned it is easy to see that the Celsius scale is 0° when the Fahrenheit is 32° so if we have a temperature in Celsius and we want to know what it is in Fahrenheit we know that we will have to add 32° to it. That will work to get from 0°C to 32°F, but it breaks down as we go up the scale, for instance 10 + 32 = 42, not 50.

Closer observation shows that the numbers on the left (the Celsius Scale) are an even 10° apart while the numbers on the right (the Fahrenheit Scale) are 18° apart. We can write this in a ratio format

$$\frac{18F}{10C} = \frac{9F}{5C}$$

The ratio gives us the rest of the conversion. If we are given a temperature in Celsius we can multiply it by the ratio then add the 32. For 20 C it would look like this:

$$20\frac{9}{5} + 32 = 68^\circ\text{F}$$

Once we know we have a working relationship we can use variables to write a general relationship that can be used without the chart for any temperature as shown below.

If F = the temperature in degrees Fahrenheit and C = the temperature in degrees Celsius

Then $F = \frac{9}{5}C + 32$.

2.2.3 Puzzling it All Out

Learning to play a musical instrument, build a canoe, or any other worthwhile endeavor requires active involvement before the information provided can be fully integrated. The study of mathematics is no different.

Work each of these puzzles on your own until you are sure you have the correct answer. Then check the answer in the back of the book. If you are having trouble agreeing with the book's answer you can: check the solutions manual if one is available in your area or ask an online program to show you a solution, ask a friend or a tutor, or ask your teacher for advice. Keep in mind that there may be more than one way to solve each problem. Always look for the easiest solution.

It is preferable that you understand how to get the correct solution for every puzzle on this page before moving to the next section.

1. What is the temperature in Fahrenheit if it is 5°C?

2. Dan runs to school at 4 mph every morning. The school is 3 miles from his home. This morning he ran into an old friend after running only 15 minutes. The friend gave him a ride the rest the way to school. What proportion of his normal ride did Dan ride this morning?

2.3 Business Formulas

How much money can I expect from an investment? What is a good price to set for my product? Which business investment will bring the largest return over the shortest time? All these questions and more are routinely answered using formulas developed for business. We will look at just two of these here; simple interest and cost.

2.3.1 Simple Interest

If we invest some money over a period of time at a particular interest rate, we will earn a return on the investment equal to the amount invested time the rate of interest time the time invested or

where R = return, P = principle (amount of money invested), r = rate, and t = time

$$R = Prt$$

If we want to know the total amount we will have at the end of that period we have to add back the amount we started with.

Letting A = the total amount we have at the end of the investment
and R = return, P = principle (amount of money invested), r = rate, and t = time

$$A = P + R = P + Prt$$

Noting that we have repetitive Ps, we can apply the distributive property to obtain a simple formula

$$A = P(1 + rt).$$

Example

Donald invested \$100,000 at 3% per annum (0.03 per year). How much money did he have from the investment after 10 years?

We can use the formula directly. First, identify the variables

$$P = \$100{,}000,\ r = \frac{0.03}{year},\ t = 10 \text{ years}$$

$$A = 100000[1 + .03(10)] = \$130{,}000$$

2.3.2 Cost

Although every item produced has a unique cost, the total cost for the production of any item is a combination of fixed costs; building rent, insurance, and other costs that are not dependent on the number of items sold and variable costs; material and other costs directly connected to item production. The total cost then can be found by adding the fixed costs to the variable cost per item times the number of items sold as shown below:

Let C = Total Cost, F = Fixed Costs, v = variable cost per item, n = number of items

Then $C = F + vn$.

Example

An art student is thinking about selling ceramic items she makes. She can buy the material to make bowls for $1 a bowl, but she has to pay $50 per hour for studio time and a $25 kiln fee for up to 100 bowls. She can make 20 bowls in an hour. If she can sell the bowls for $5 each can she make a profit?

Since we are within the limits on the fixed costs, we can use them directly. Therefore, the fixed costs are $F = 50 + 25 = 75$. We can then apply the formula directly to obtain the total cost:

$$C = F + vn = 75 + 1(20) = \$95.$$

We find the gross (before costs) amount of money she will make by multiplying the selling price by the number of items

$$5(20) = \$100.$$

Since the cost is higher than the selling price she will make a profit ($5).

2.3.3 Puzzling it All Out

Learning to play a musical instrument, build a canoe, or any other worthwhile endeavor requires active involvement before the information provided can be fully integrated. The study of mathematics is no different.

Work each of these puzzles on your own until you are sure you have the correct answer. Then check the answer in the back of the book. If you are having trouble agreeing with the book's answer you can: check the solutions manual if one is available in your area or ask an online program to show you a solution, ask a friend or a tutor, or ask your teacher for advice. Keep in mind that there may be more than one way to solve each problem. Always look for the easiest solution.

It is preferable that you understand how to get the correct solution for every puzzle on this page before moving to the next section.

1. What would be the total amount from the example in 2.3.1 if the rate of interest was 0.25% per month over the same 10 years?

2. What is the total cost to manufacture 10,000 tennis balls at a variable cost of 10 cents per ball and a fixed cost of $300?

2.4 Chapter 2 - Review

In this chapter we looked out how we use variables to develop formulas that can help us solve problems quickly and easily. We learned that algebraic formulas are used in nearly every field and study. We looked at how a few geometric, scientific, and business formulas were developed and how they could be used to help solve problems. We found that once a formula is developed for a given problem, we can simply substitute numbers in for the appropriate variables and evaluate the expression to obtain a solution.

2.4.1 Geometric Problems

We looked at geometric formulas and found that:

$$P = L + W + L + W \quad \text{and} \quad A = LW$$

Where P = Perimeter, A = Area, L = Length and W = Width (for a rectangle),

$$A_T = \frac{1}{2}bh$$

where A_T = Area, b = base, and h = height (for a triangle), and

$$C = \pi d \quad \& \quad A = \pi r^2$$

where C = the circumference, A = the area, d = the diameter, and $r = \frac{d}{2}$ or $\frac{1}{2}$ the diameter (for a circle).

2.4.2 Scientific Problems

We look at the Physics formula used to predict the distance at object moves given a rate of speed and a time:

$$d = rt$$

where

d = the distance covered, r = the rate (or speed) the object is traveling, and t = the time the object is moving at the given rate.

And we developed a formula that converts the temperature in degrees Celsius to the temperature in degrees Fahrenheit:

$$F = \frac{9}{5}C + 32$$

Where F = the temperature in Fahrenheit and C = the temperature in Celsius

2.4.3 Business Formulas

We developed and used formulas to calculate the total amount of money we will have after an investment returning simple interest:

$$A = P(1 + rt)$$

where A = the total amount at the end of the investment
P = principle, r = rate, and t = time.

Finally, we looked at how we might calculate the cost of producing something to sell and developed the formula:

$$C = F + vn$$

where C = Total Cost, F = Fixed Costs,
v = variable cost per item, n = number of items.

2.4.4 More Practice Puzzles

Remember to define your variables when writing formulas.

1. The formula for the surface area of a cube is

$$A_{sc} = 6l^2$$

Where A_{sc} is the surface area of a cube and l is the length of one of its sides.

Find the area of a cube with 6 inch long sides, compare this to the surface area of a cube with 4 inch sides using a rational form.

2. Tanya spent $20 on equipment rental to make a batch of bracelets. Her materials will cost her $5 per bracelet. Use the cost formula to calculate how much money it will cost her to make 120 bracelets.

3. John's parents started a bank account for him with $100. He plans to put $50 a month from his mowing job into the account to save for a new mower that costs $320. Neglecting any interest earned, develop a formula that will help John predict how much money he will have after any particular time. Use the formula to find out how much money John will have saved after 2 years. Will he have enough for the mower?

4. Write a formula for the area of the triangle in the diagram using only one variable.

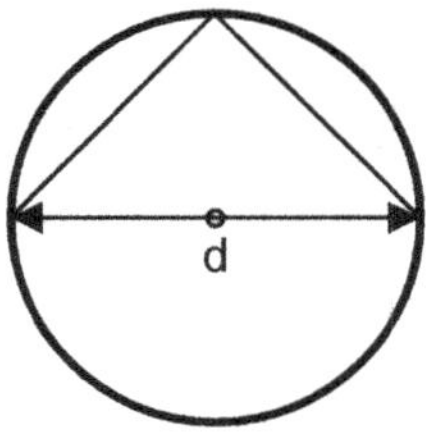

5. The number of a certain bacteria after growing them over a period of time may be modeled by the formula

$$N = N_o + N_o(1+r)^t$$

Where N = the number of organisms after a certain period of time,
N_o = the original number of organisms,
r = the rate of growth, and t = the certain period of time.

Two different lab teams run experiments using environments supporting different growth rates. One team enriches the environment and achieves a growth rate of growth of r = 30 per hour. The other team holds back a vital nutrient and achieves only r = 5 per hour. How many organisms does each team expect after 4 hours growth assuming they each start with a culture containing 10 bacteria?

6. Simple algebraic manipulation (which we will learn in Unit 2) can be used to transpose the distance formula into a formula that compute the time required to go a particular distance at a certain rate as:

$$t = \frac{d}{r}$$

Where t = travel time, d = distance, and r = rate.

I want to make a trip to Salem to pick up a set of curtains I ordered but I don't want to drive at night. The maximum daylight hours available to me is six. If Salem is 160 miles from my home and the fastest I can travel is 40 miles per hour, use the formula to tell me if I should try and make the trip?

Unit 2 – Exploring Relationships

Probably the most powerful use of variables in mathematics is exploring relationships. We have many ways of communicating connections between things. In this unit we will see how the development of some very structured mathematical ways of looking at relationships has allowed amazing understanding and manipulation of our world.

Chapter 3 – Defining Mathematical Relationships

3.1 – Picture a Relationship…

As we saw in Chapter 2, we can define a mathematical relationship between two or more varying quantities using a formula. Most of us have used pictures and graphs at some time to try and to explain how the change in one thing can affect another. This chapter is a short breath to let you think about the many ways that we convey this information to each other and use it to make decisions and solve problems in our lives. The best way to perform the activity below is with a group. If this is not possible for you, do them anyway trying to be as creative as possible.

[Note - The same neurons we need for mathematics are exercised when we use creative thinking. Adding as much creativity and art to mathematical solutions enhances the learning experience.]

3.1.1 Class Activity

Your class has been selected to evaluate the performance of the Willy Wonka Widget Works (WWWW) in various areas. The company makes widgets and sells them all over the world. The class should divide into teams. Each team should create their own data and develop a presentation to show the performance in each of one or two areas. Each team then presents their ideas at a group meeting. The class should decide, based on the presentations, whether WWWW is performing well enough to receive the funding they are requesting or, if not, make suggestions for improving their business.

Teams may select from the following areas and/or create their own:

Profit/Loss and Time
Product Cost and Profit
Items Sold and Time
Business Expenses and Profit
Product Cost and Time
Total Profit and Number of Items Sold
Total Cost and Number of Items Produced

There are no wrong answers and presentation counts!

3.2 – René Descartes and the Cartesian Coordinate System

As you can see from the Chapter 3 Activity, there are many ways to picture a relationship. Illustrations and charts, pie graphs and columns, mathematical formulas, and other tools give us insight and information for the relationship in question but it may often be difficult to compare this information to that for a different relationship.

In the early 17th century a French philosopher named René Descartes developed a system that allows us to quickly portray information about a relationship in a standardized manner that can be easily and accurately compared with that for similar relationships. This simple invention allowed the development of analytical geometry and calculus. Science could not have emerged as it did in the subsequent years without these important mathematical tools. For this reason, every modern day convenience we now take for granted owes its existence to René Descartes.

As you will see in this and the subsequent chapters, the Cartesian coordinate system that Descartes developed is an important tool in its own right.

3.2.1 Coordinate Axes

The basis of the Cartesian coordinate system is a set of three infinite lines intersecting at a single point perpendicular to (at a right angle with) one another to represent three dimensional space. You might think of it as a small camera view finder that can be placed at any point of interest to give a close up picture of the immediately surrounding area. We can zoom as close or as far away as we like to try and capture the right image. Sometimes, we might take many different pictures, preserving both detail and overall concept.

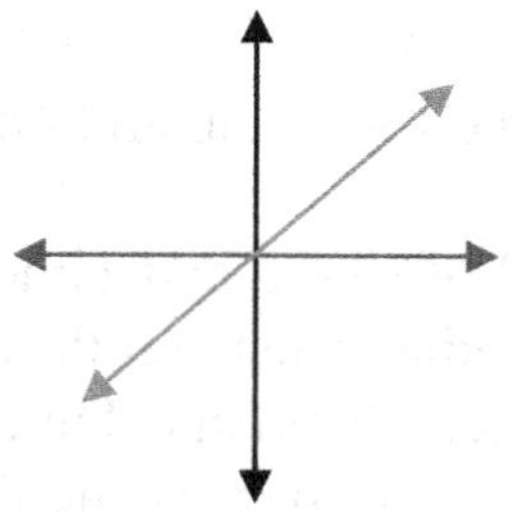

The Cartesian Coordinate Axis

With the Cartesian coordinate system, we draw the axis and the information we wish to look at in a very prescribed manner. We can then glean specific information from our picture quickly and easily. If we set our "zoom" the same (draw both axes to identical scales) we can quickly compare one situation to the other. In addition, as previously mentioned, the system has given us a way to picture mathematical relationships, a somewhat abstract concept, in a very real way leading to inspiration and development in every corner of life.

To simplify our learning of the Cartesian coordinate axis, we will limit our discussion to a set of two intersecting axes. Think of this as the picture you get when you take the photo. Although the camera was aimed through three dimensional space, the photo is a flat, two dimensional, object. Although it really only represents the single surface where the camera was focused, the photo, the two dimensional representation of the three dimensional space, still gives us significant information. Keep in mind as we work through this book that any of the practices and properties used for two dimensional Cartesian coordinate analysis can be extended for use in a three dimensional system.

3.2.2 – Points and Quadrants

The two dimensional Cartesian coordinate axes consists of two infinite intersecting perpendicular lines. The first thing one might notice about arrangement is that it divides space into four areas. Mathematicians have named these areas Quadrants (since there are four of them) and assigned them Roman Numerals to identify each. The labeled Quadrants are shown on the axes below:

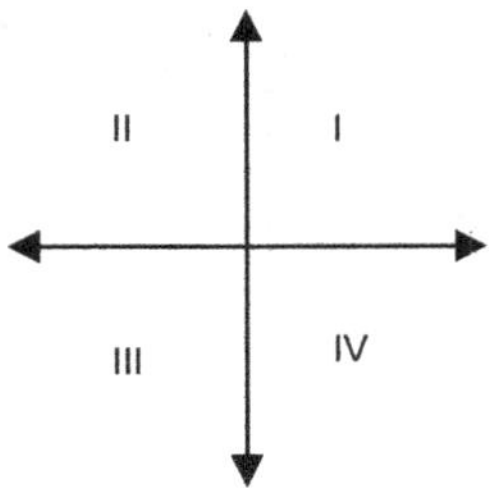

Each axis can be labeled to represent any unknown quantity in a relationship that we want to explore. Obviously, when we use a two dimensional analysis, we can only explore relationships with two unknown variables. We can also divide each axis into any scale that suits our needs. Often, we just use x and y as generic variables and divide each axis with identical scales and the negative values to the bottom and on the left as shown below:

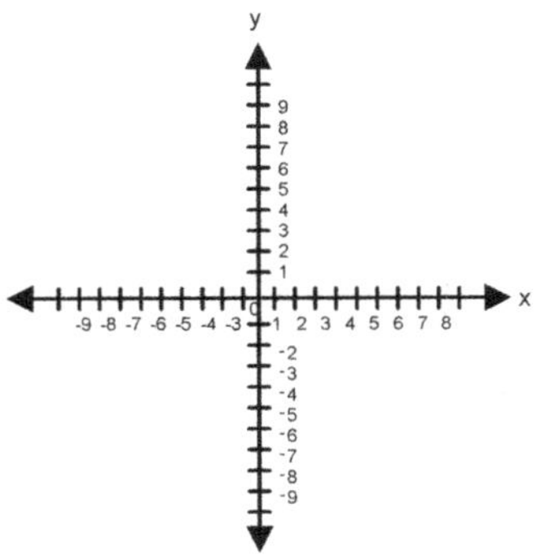

A Typically labeled Cartesian Coordinate Axes

Once we have a set of axes set up and labeled, we can find points in space. The consistency of the location of these points in space on a specifically labeled set of axes gives us the ability to compare.

In order to locate a specific point in space we must have a value for each variable. The values are then found by locating the intersection between the value of one variable (on its axis) and the value of the other in its. An example of the mathematical syntax for a point is (x,y) where x represents the variable on the horizontal axis and y represents the variable on the vertical axis. A few points are plotted and labeled on the axis below:

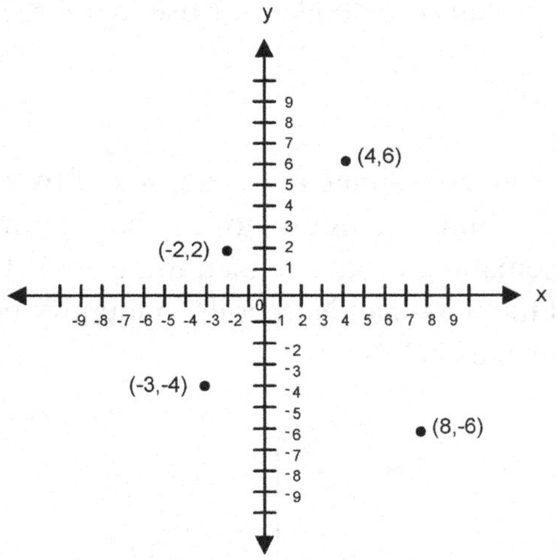

3.2.3 – Connecting the Dots

It is hard to see a relationship between the variables in the points plotted in Section 2. We refer to such an arrangement as a random point distribution. We could connect the dots in any order and our picture would still be random. Often, however, when we connect the points in a relationship we observe very unique patterns and forms. Many of these patterns and forms have been recognized and developed into useful mathematical tools.

3.2.4 – Puzzling it All Out

Learning to play a musical instrument, build a canoe, or any other worthwhile endeavor requires active involvement before the information provided can be fully integrated. The study of mathematics is no different.

Work each of these puzzles on your own until you are sure you have the correct answer. Then check the answer in the back of the book. If you are having trouble agreeing with the book's answer you can: check the solutions manual if one is available in your area or ask an online program to show you a solution, ask a friend or a tutor, or ask your teacher for advice. Keep in mind that there may be more than one way to solve each problem. Always look for the easiest solution.

It is preferable that you understand how to get the correct solution for every puzzle on this page before moving to the next section.

Get some practice plotting points and recognizing patterns with the following activities:

1 Draw a two dimensional set of Cartesian coordinate axes on a sheet of graph paper. Use equal scales for each axis with intervals that will allow you to plot and number the following points. Plot and number the points. Then connect the dots to reveal a hidden treasure.

1	(0,-13)	8	(-2.5,1)	15	(0,8)	22	(2.5,1)
2	(-0.25,-12.5)	9	(-2.5,4)	16	(0.5,7.5)	23	(2.25,0)
3	(-0.5,-12)	10	(-2.25,5)	17	(0.5,1)	24	(1.5,-0.5)
4	(-0.5,-1)	11	(-2,4.5)	18	(2,1)	25	(0.5,-1)
5	(-1,-0.5)	12	(-2,1)	19	(2,4)	26	(0.5,-12.5)
6	(-2,0)	13	(-0.5,1)	20	(2.25,5)	27	(0,-13)
7	(-2,25,0.5)	14	(-0.5,7.5)	21	(2.5,4.5)		

2 Plot the following set of points on a two dimensional axes and connect the dots.

(-11,-33)
(-8,-24)
(-3,-9)
(1,3)
(6,18)
(7,21)

What is different about this pattern than the pattern produced in puzzle 1?

Food for Thought

Can you write a mathematical formula to express the relationship you see?

Chapter 4 – Linear Relationships in the Cartesian Coordinate System

In Chapter 1 we were reminded that when multiply two times three we get 6 and that we could write the relationship as:

$$2(3) = 6.$$

Although this is a specific, constant, relationship, it represents a family of relationships that could be described as "what happens when I multiply something by three". To explore this relationship in the Cartesian coordinate system we must write all the other numbers but the 3 in terms of variables as shown below:

$$y = 3x$$

where x an y are any real numbers.

4.1 – Charting a Line

4.1.1 – Graphing a Linear Relationship

We can use the equation $y = 3x$ to obtain any number of points to plot on a Cartesian axes. Choose any x and use the equation to calculate the y as shown below:

x, y

(0, 0)
(2, 6)
(-2, -6).

The plotted points are shown below:

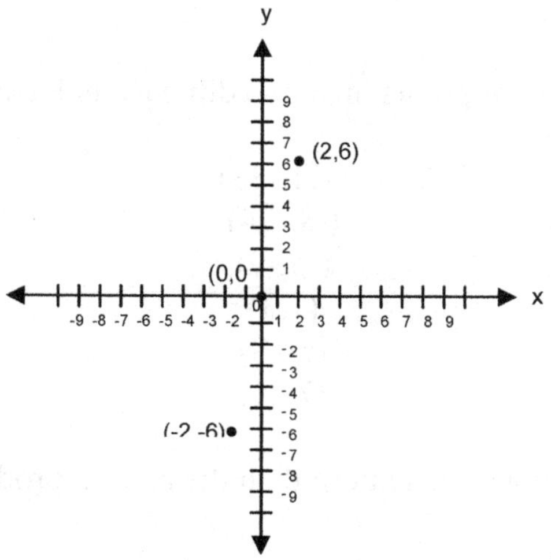

It is easy to see that a line drawn to connect the dots will be a straight line as shown below. Each student is encouraged to calculate and plot as many points as is necessary to fully see this.

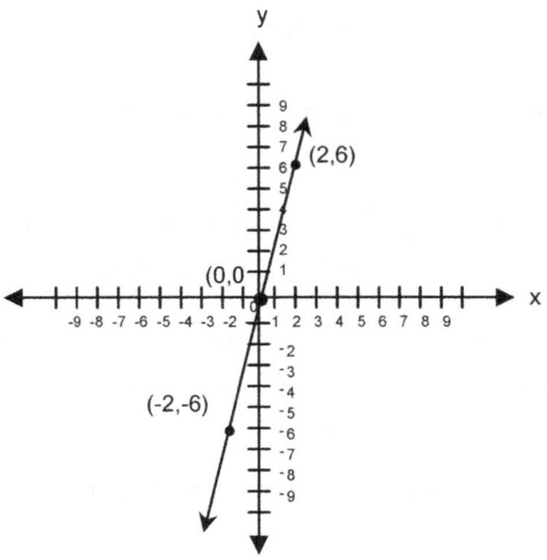

Although it might seem a little abstract to see a picture of a mathematical relationship on a graph, the visual can assist us in many ways. For instance, can you look at the graph we just made and tell what 1 times 3 is without calculating it? It should be easy to see that you need only find 1 on the x axis and follow it vertically until it intersects the line. In this case we will get 3. We can validate the answer through calculation if desired. (Try this with other values.)

We can also use the system to easily compare relationships. Let's graph another simple linear relationship next to the one we just graphed. We will first develop a chart of points as before,

x,	y
(0,	0)
(4,	2)
(-6,	-3).

We can then graph the points and connect them with a line. To compare the two relationships we can draw both on the same set of axes as shown below:

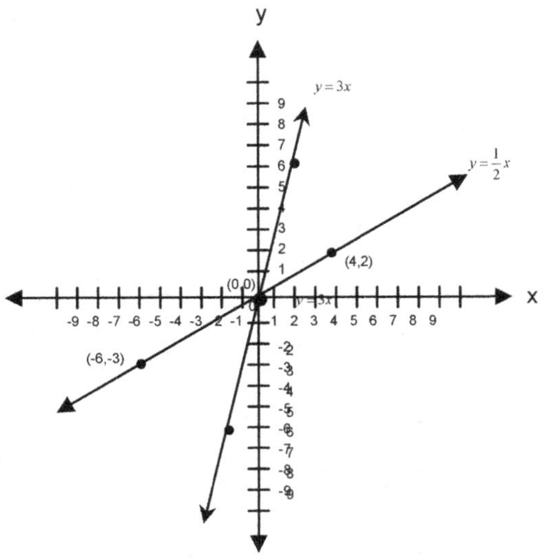

There are a few things that we notice right away:

Both relationships form a straight line.
(We call these linear relationships.)

Both relationships cross each axis at (0,0).
(We call this point the origin)

The only real visible difference between the two relationships is the amount that one tilts or is rotated with respect to the other.
(We call this property the slope.)

The line $y = 3x$ has a steeper slope that the line $y = \frac{1}{2}x$.

It would be desirable to use these observations to develop a general formula to mathematically represent a line in the Cartesian coordinate system.

4.1.2 – Slope

One clue to this is in the obvious difference between our two lines, the tilt or rotation. We can easily see that for these two lines, the amount of rotation appears to be reflected in the constant that appears in each formula as the line with 3 in its formula is significantly steeper that the line with only $\frac{1}{2}$. A clue to the reason for this lies in the operation being performed in the formula and the chart that we used to draw the line. If we choose more evenly dispersed x values to generate the y values and look a little more closely we will quickly see a direct relationship.

Comparison y = 3x and y = 0.5x				
x value	3x		1.5x	
		change		change
-5	-15	3	-2.5	0.5
-4	-12		-2	
-3	-9	3	-1.5	0.5
-2	-6		-1	
-1	-3	3	-0.5	0.5
0	0		0	
1	3	3	0.5	0.5
2	6		1	
3	9	3	1.5	0.5
4	12		2	

The amount of tilt can be described as the amount of change between each of the points that we plot for the line. It is clear that for a line (or linear relationship) this change is the same between any two points and exactly the same as the amount we are multiplying by.

As stated before mathematicians have defined this change as the slope of the line and assigned it the variable m. Using this definition, we can easily write a general formula to describe similar lines in a two dimensional Cartesian coordinate system as shown.

A line crossing the origin in a 2D Cartesian system can be represented as:

$$y = mx$$

where x and y are any real numbers and

m equals the slope of the line ($\left(\dfrac{\Delta y}{\Delta x}\right)$, the constant change that produces it)

The constant multiplicand in a linear relationship is the slope of the line when we draw it in the Cartesian system. We often call this constant the coefficient of the variable.

4.1.3 – Intercepts

Unfortunately, there are many lines that can be drawn that do not have a point at the origin. If we want a truly general formula then, we must start with a more general relationship than those just considered. We will start with a picture and, based on what we have observed so far, try and develop a more useful formula

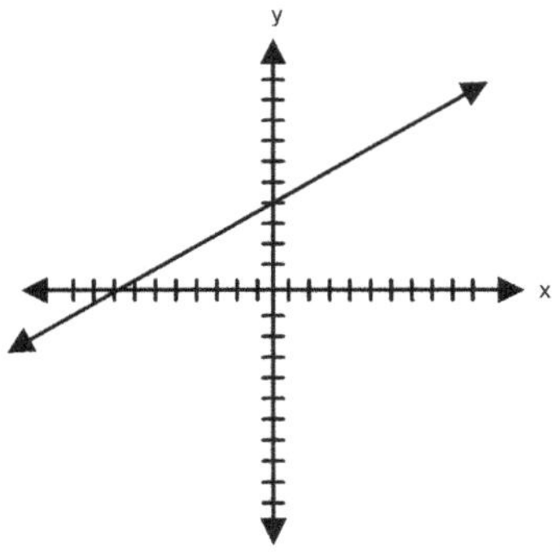

When we move the line from the origin to make it more general, we notice that it still crosses each axis but now in separate places. We might observe that, except in the case of crossing at the origin, any line that might be drawn will cross each axis in a unique location. [Draw a few more lines until you see this for yourself].

We call the point where a line crosses an axis an intercept. As observed, a two dimensional set of Cartesian coordinate axes will have two intercepts, one for each axis. Since these two intercepts are obviously unique for each line, labeling them might facilitate the development of a formula to describe the line (and predict values along it). We have done so below:

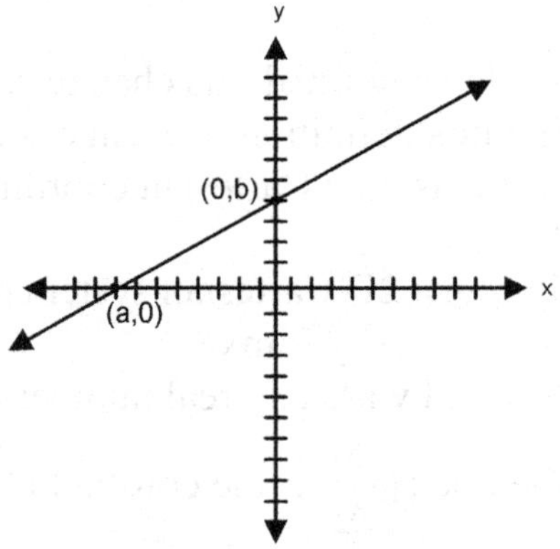

(where a and b are any real numbers)

When we label the points, it becomes obvious that each intercept crosses its axis at the zero value of the other axis. In our diagram the x intercept crosses where x = a and y = 0. The y intercept crosses where x = 0 and y = b.

4.1.4 – Slope Intercept Formula for a Line

We might also observe that we could move the line to pass through the origin without changing its slope so we can still represent that property of the line with the formula previously developed. In addition, one intercept and the slope are all that are required to define the line for once these are set, the other intercept is inevitable as shown in the following example:

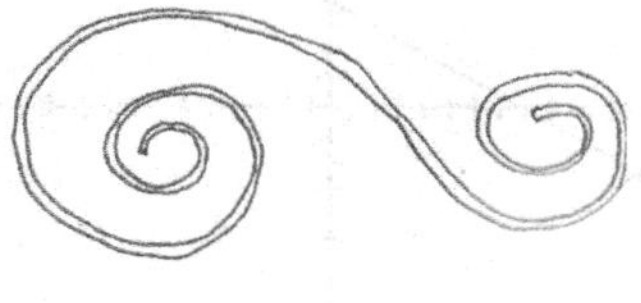

Example

A line with the a slope of $m = \frac{1}{2}$ *intercepts the y axis at (0,4).*

Where will it intercept the x axis?

We can draw an illustration of the question using Cartesian coordinates. The y intercept (where x = 0) is set at (0,4). We will let a = the x value of the x intercept as shown below.

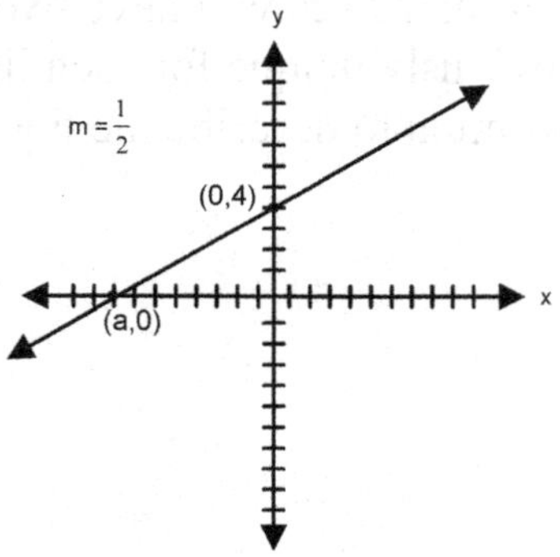

We can readily see that if the value in the y direction changes by 4 from the origin, the value of a must be 8 if the slope of $\frac{1}{2}$ is to be maintained for we know that $\frac{1}{2}=\frac{4}{8}$. We can draw another illustration to show this:

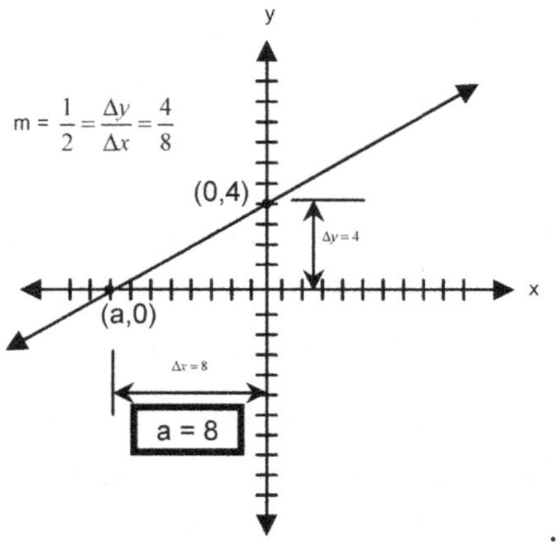

The line will cross the x axis at x = 8 and y = 2.

[Note – The following discussion uses a format called function format often used in higher mathematics when describing relationships. In this format, $y(x) = mx + b$ is read "y, a function of x, is equal to mx + b". The word function is just a fancy way of saying that when one variable changes, the other will change in a predictable way (according to a rule or "function"). A particular point on a line, say the y intercept in the preceding example, is represented in function format by the symbolism $y(0) = 4$, read the value of a specific function when x equals 0 is 4.

Unfortunately, students in early algebra are also just learning the $y(-6)$ format for multiplication and function format can be confusing. Because of this, it is rarely used and often not introduced at this level. It is included here to familiarize the student with the format in a discussion where the meaning will be readily obvious.]

If we look simultaneously at the line in the above example with a line of the same slope going through the origin as shown below, we will find a pattern that will help us develop our general formula. We will start by looking at the difference between the values for each line at x values marked with dashed lines, 4 and -6.

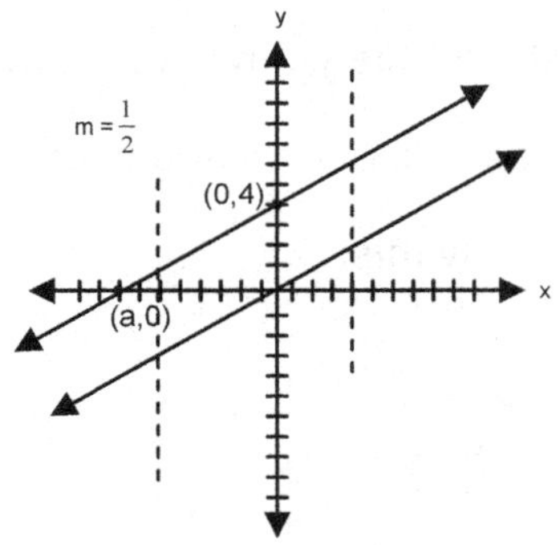

We already know the formula for the line with intercept (0,4), $y = \frac{1}{2}x$. For the line through the origin we can calculate the value of y when x = 4 as 2 since $\frac{1}{2}(4) = 2$. We can read the value of y when x = 4 for the other line from the graph. It is 6. We will note here that $y_4(4) = 6$ (the y value at x = 4 from line intercepting the y axis at 4) minus $y_0(4) = 2$ (the y value at x = 4 from line intercepting the y axis at the origin) equals 4, the y value of the x intercept. A similar analysis at x = -6 yields

$$y_0(-6) = \frac{1}{2}(-6) = -3.$$

From the chart we find $y_4(-6) = 1$. And again, $y_4(-6) - y_0(-6) = 1 - (-3) = 4$.

It would appear that no matter where we choose to investigate, we can define the line intercepting the y axis at 4 completely by using the formula for the line passing through the origin and adding 4 (the y value of the intercept) as shown below.

$$y = \frac{1}{2}(x) + 4$$

Indeed, we can immediately see that an x value of zero yields the point (0,4) that we used to define our line. Chart this new formula and you will find that you have reproduced the line with a slope of 4 and a y intercept at (0,4).

This gives us confidence as it is easy to see that the formula can quickly be generalized to account for a line that crosses any intercept by merely using a variable to represent which intercept it is crossing. We generally use the variable b to represent the y value of the y intercept and with it we can write a general equation for any line in a 2D Cartesian coordinate system as

$$y = mx + b$$

represents a line in a 2D Cartesian Coordinate System

where x and y are any real numbers,

m = the slope of the line $\left(\frac{\Delta y}{\Delta x}\right)$,

and b = the y value of the y intercept of the line.

The $y = mx + b$ form is often called the slope intercept form of the equation of a line.

4.1.5 Using the Graph of a Linear Relationship to find solutions

In the above discussion we read values from the graph for our function by finding the value where the line crosses the y axis at a given value for x. We knew that we could do this because we knew the line to be a connection between points each with a unique value of y for any x. As we will see in the next section, each of these points is also an intersection between the line representing the function and the line representing $y = d$ where d equals the value of y at the point we are looking for. This observation leads to one of the most powerful tools that algebra, and analytic geometry, has to offer.

4.1.6 Puzzling it All Out

Learning to play a musical instrument, build a canoe, or any other worthwhile endeavor requires active involvement before the information provided can be fully integrated. The study of mathematics is no different.

Work each of these puzzles on your own until you are sure you have the correct answer. Then check the answer in the back of the book. If you are having trouble agreeing with the book's answer you can: check the solutions manual if one is available in your area or ask an online program to show you a solution, ask a friend or a tutor, or ask your teacher for advice. Keep in mind that there may be more than one way to solve each problem. Always look for the easiest solution.

It is preferable that you understand how to get the correct solution for every puzzle on this page before moving to the next section.

1. Find the length of the line labeled c using Pythagorean's Formula ($c = \sqrt{a^2 + b^2}$ where a and b are equal to the length of each perpendicular side of a right triangle and c is equal to the hypotenuse or longest side). Hint – the x axis is always perpendicular to the y axis.

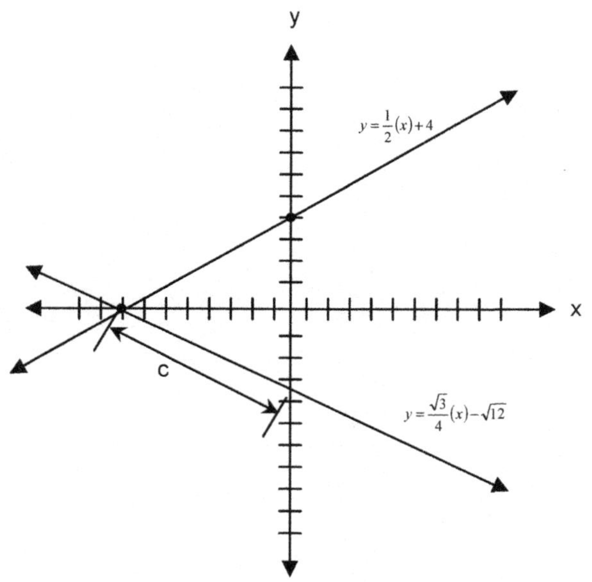

2. Graph the lines represented by $y = 2x + 6$ and $y = -3x + 6$.
   Find the area of the triangle formed between these two lines and the x axis (y = 0) using the formula for the area of a triangle from Chapter 1 or the back cover from your book.

4.2 – Graphing and Interpreting Linear Relationships

4.2.1 – Using the Slope Intercept Form of an Equation to Graph a Line

It should be obvious from the discussion in Section 5.1 that any linear equation that is in the y = mx + b format can be readily graphed on a set of Cartesian Axes without the use of charts and calculations as shown in the example below.

Example

Graph the relationship y = 3x + 4.

If we compare the equation with the general linear form we can immediately identify the slope (m) as 3 and the y intercept (b) as 4 as shown below

$$\begin{array}{c} y = mx + b \\ \quad\;\;\downarrow \quad \downarrow \\ y = 3x + 4 \end{array}$$

The y intercept gives us one point (0,4) which we can readily plot on a 2D set of axes.

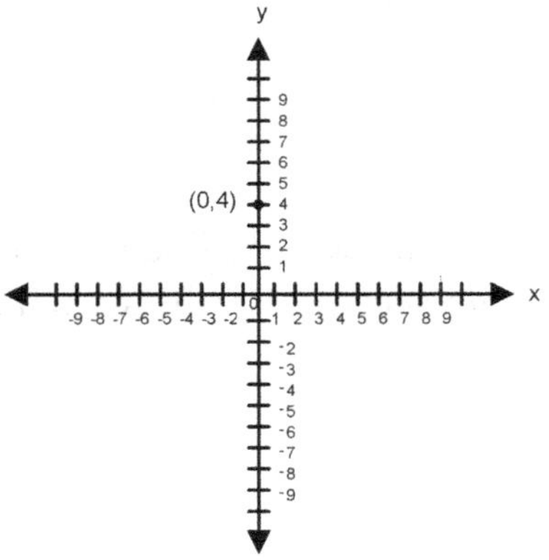

If we write the slope in rational form and use its definition (the change in y over the change in x $\left(\frac{\Delta y}{\Delta x}\right)$ we can use a similar comparison to find another point.

$$\left(\frac{\Delta y}{\Delta x}\right) \rightarrow \left(\frac{3}{1}\right)$$

We note from our comparison that for every interval we move from the line in the x direction, we must move 3 intervals in the y direction to return to the line. We can use this relationship to plot several more points as shown.

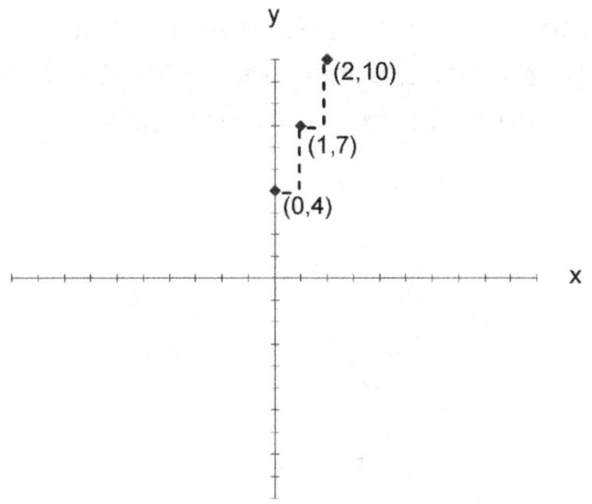

With at least 2 points, we can draw the line that connects them.

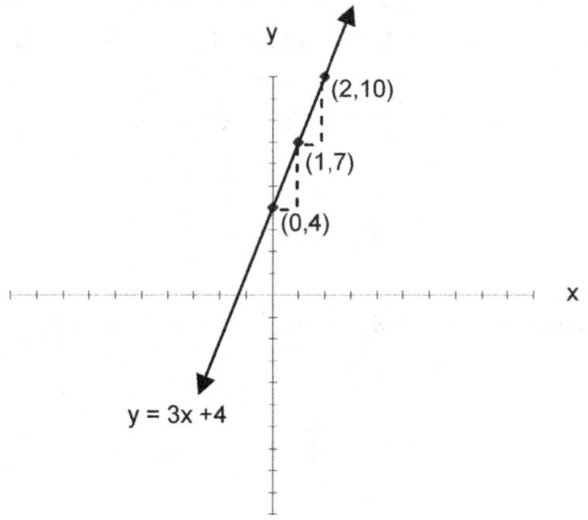

Once we have drawn a graph, we can use it to find other points along the line directly. For instance, the point that intersects y = 0 (the x axis) is approximately (0,-1.3).

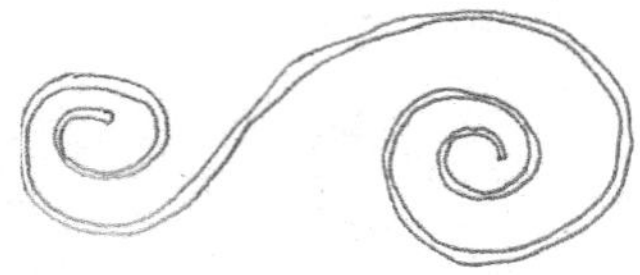

We can use the general formula even if it contains negative values by first applying the rules for signs. An example showing this is given below.

Example

Graph the relationship $y = -2x - 3$.

To compare with the general form as before we must first rewrite the equation so that we can directly extract the slope and the y intercept. This is easy to do by applying the identity property for multiplication and division and noting that

$-2 = \frac{-2}{1} = \frac{2}{-1}$ and that subtracting 3 is the same as adding 1(-3). Note than we can use a negative sign in either the divisor or the dividend to maintain the negative quotient.

$$y = mx + b$$

$$\downarrow \quad \downarrow$$

$$y = \frac{-2}{1}x + 1(-3)$$

Again the y intercept gives us one point (0,-3), this time falling below the x axis.

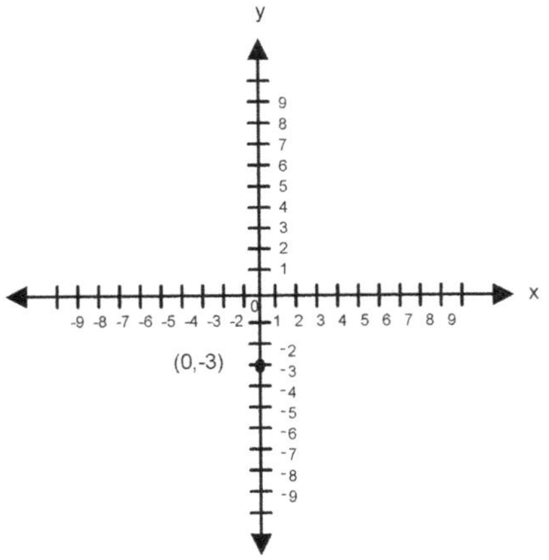

The slope is already in rational form and we can extract directly

$$\left(\frac{\Delta y}{\Delta x}\right) \begin{matrix} \rightarrow \\ \rightarrow \end{matrix} \left(\frac{-2}{1}\right)$$

We can now use the slope to plot several more points and connect them as before.

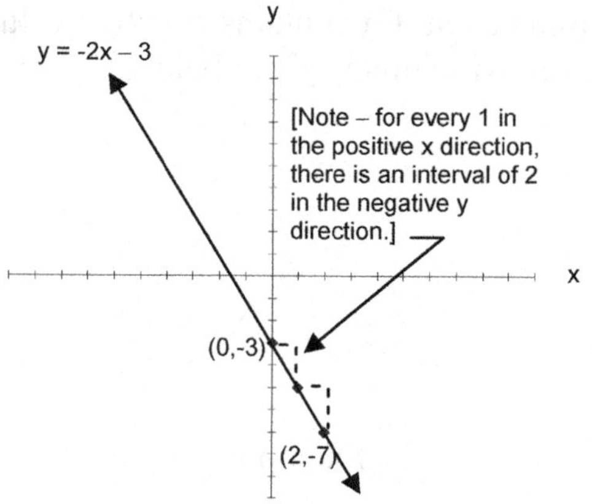

4.2.2 – Common Linear Graphs and Interpretations

The preceding examples illustrate an important property of linear relationships graphed in Cartesian coordinates. Equations with a positive coefficient slope to the right while those with a negative coefficient slope to the left. Knowing this can help us use Cartesian graphs to make comparisons quickly and easily.

Example

Performance for the last year was studied for Company A and Company B. A graph showing their profits over time for the previous year is shown on the next page. Choose which company would be a better investment for your portfolio and tell why.

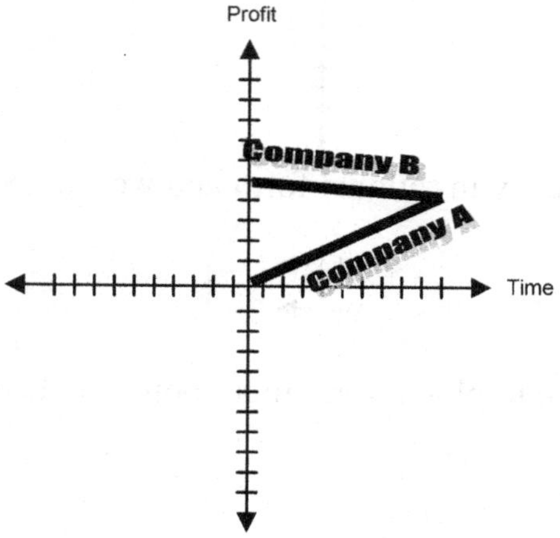

Without values we can still see clearly that Company A began as a new company or just breaking even at the beginning of the year and has steadily produced higher profits throughout the year. Company B, on the other hand, started out making higher profits but profits have steadily declined. Unless Company B can show a good reason for its declining profits, Company A is a better portfolio choice.

We can adapt the y = mx + b format to write equations for horizontal lines in Cartesian coordinate systems also. A horizontal line, as shown below, has a y intercept, but there is no apparent slope.

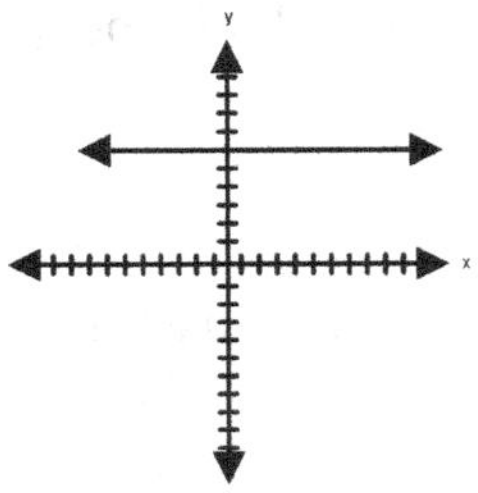

Using the definition for slope $m = \frac{\Delta y}{\Delta x}$, we can see that for 1 interval of change in the x direction, there is a 0 change in the y direction. We can write this as $m = \frac{0}{1} = 0$. Keeping the y intercept general we can make the substitution for m in the general equation and simplify yielding

y = 0x + b   or   y = b
General Equation for a Horizontal Line
(where x and y are any real numbers and b is the y value of the y intercept)

A vertical line presents more challenge

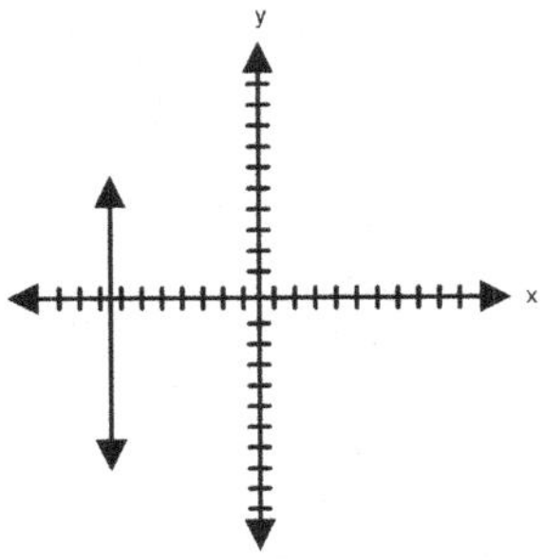

There is no y intercept and if we explore the slope as we did for the horizontal line we will find that for any movement of one interval in y, there is a corresponding 0 interval in x or,

$$m = \frac{1}{0}.$$

Since we can obviously not divide anything into zero groups, this expression is meaningless, and we find that we have neither a slope nor a y intercept for a vertical line. However, the equation that we arrived at for a horizontal line, y = b, suggests a solution.

$$x = a$$

Equation for a vertical line

(where x = any real number and a is the x value of the x intercept)

Graph a few of these using any value for a to assure yourself that this general equation works for all cases.

4.2.3 – Parallel and Perpendicular Lines

Two unique linear forms that sometimes occur when comparing relationships are parallel and perpendicular lines. There are several observations that we may make that can enhance our understanding these relationships.

Parallel lines are lines that never cross. The distance between them is constant no matter where it is measured. When we graph parallel lines in the Cartesian coordinate system, as done below, we can see immediately that they will always have the same slope and different intercepts.

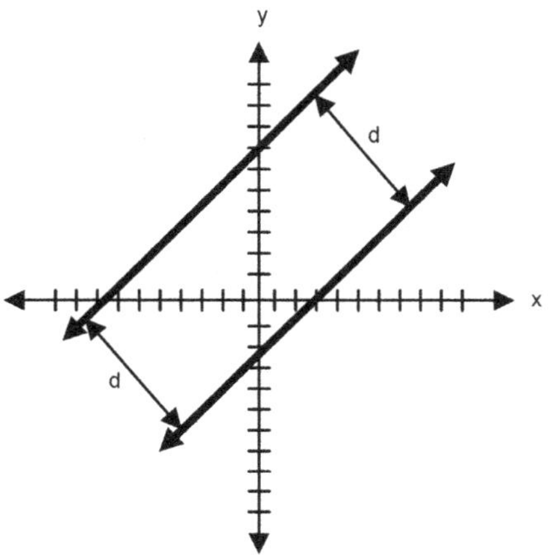

If we know that two lines are parallel, we know the slope of both of them even if only one is given.

Example

Healthy World Company is trying to develop a dosage chart for adults for a new prescription drug that is very sensitive to body mass. A chart has already been developed for children. The effects are the same per pound for both adults and children but that the children require a lower overall amount to account for their faster metabolism. The children's dosage chart has a slope of 0.25 ml per lb of body weight. The adult dosage for a 115 lb person is 50 ml. Develop a dosage chart that will give the correct dosage for adults from 100 lb to 250 lb.

We start by noting that we are given one point for the chart, 115 lbs and 25 ml or (115,50). We can easily plot this point of a set of coordinate axes. Attention to both axis labeling and scaling will help develop a useful chart as shown.

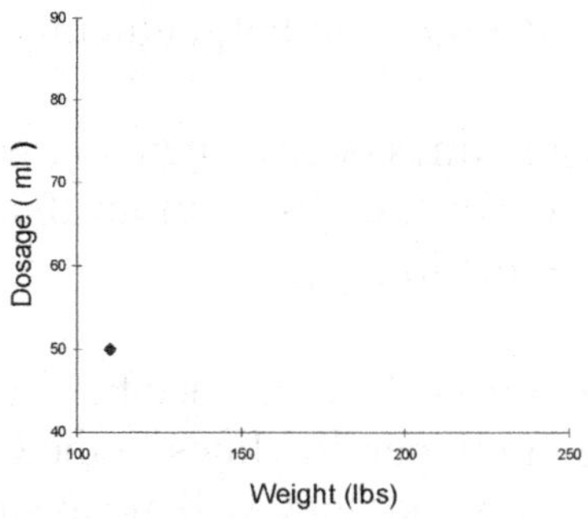

We can find another point or two from the slope as before. To avoid dealing with the decimal, we multiply by the rational form of one $\frac{20}{20}$ to find the increase in dosage per 20 pounds of weight and use this ratio to find two more points as shown below.

$$m = \frac{0.25ml}{1lb}\left(\frac{20}{20}\right) = \frac{5ml}{20lbs}$$

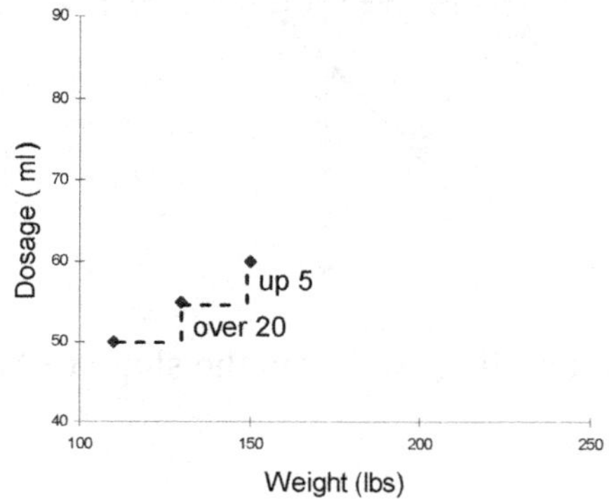

Once we have a sufficient amount of points we can connect them and extend the line.

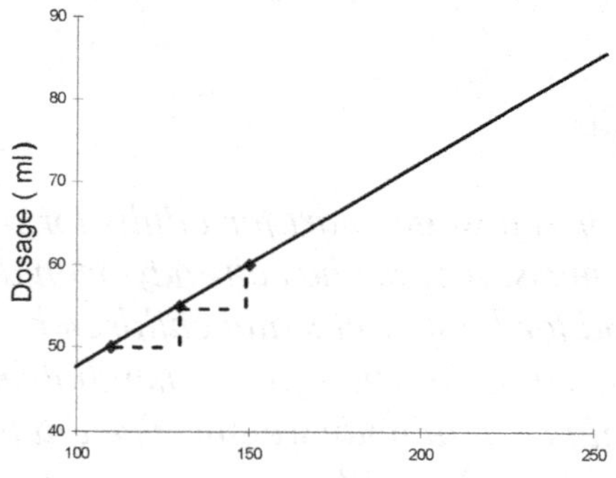

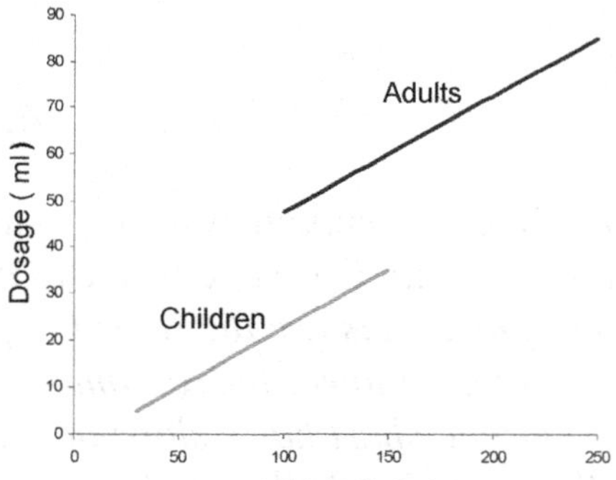

The chart may now used to find the adult dosage at any weight. A graph including the children's chart verifies that the slopes are indeed the same and the lines are parallel.

Perpendicular lines have been previously defined as lines that intersect at a 90° angle. Since the Cartesian coordinate axes is also orthogonal (meets at 90°), there is a unique relationship between the slopes of perpendicular lines. The secret lies in the perpendicularity.

The diagram below shows a set of perpendicular lines with the slope shown for one of them. The labeled slope changes a negative 3 in the y direction for a positive 6 in the x for a slope of $m = \frac{\Delta y}{\Delta x} = \frac{-3}{6} = -\frac{1}{2}$.

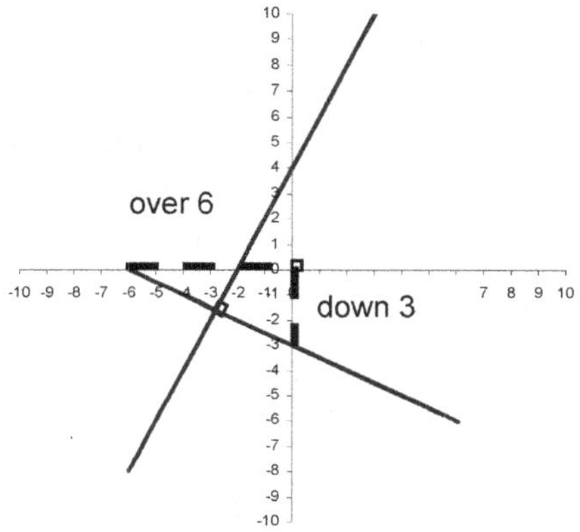

If the orthogonal lines used to show the slope are rotated 90°, we obtain the slope for the intersecting perpendicular line of $m = \frac{\Delta y}{\Delta x} = \frac{6}{3} = 2$ as shown below.

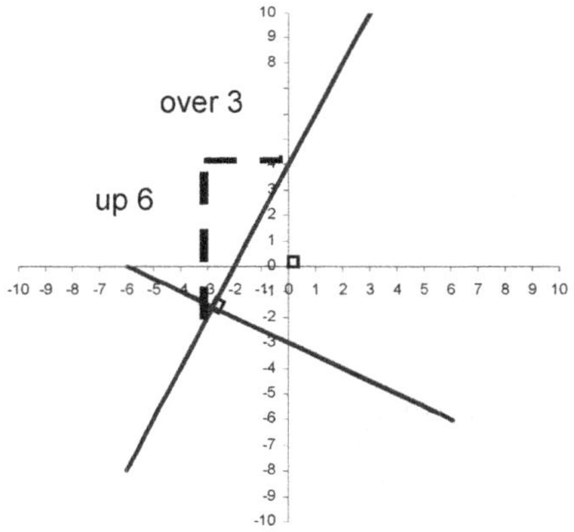

It should be obvious that the slope of one perpendicular line will always be the negative inverse (or reciprocal) of the other.

We can summarize these two relationships as follows:

$m_1 = m_2$	Parallel Lines in Cartesian Coordinates
$m_1 = -\frac{1}{m_2}$	Perpendicular Lines in Cartesian Coordinates

4.2.4 – Solutions Between Linear Relationships

Any time two lines intersect on a Cartesian set of axes, they will share common values at the point of intersection. To see this more clearly we will chart 2 relationships and see where they intersect. First we make a chart of each relationship at several different points. If the points we choose include the intersection we will see that at that point each equation returns the same y value at a given x value. For our example below, both functions return a value of 0 at x = -2.

Independent Variable (x)	y = 2x + 4	y = -x – 2
-5	-6	3
-4	-4	2
-3	-2	1
-2	**0**	**0**
-1	2	-1
0	4	-2
1	6	-3

We can graph each relationship using y = mx + b or plotting any two of the calculated points above. Usually a third point is plotted as a check. When graphed the two lines clearly intersect at the point (-2,0) as seen below.

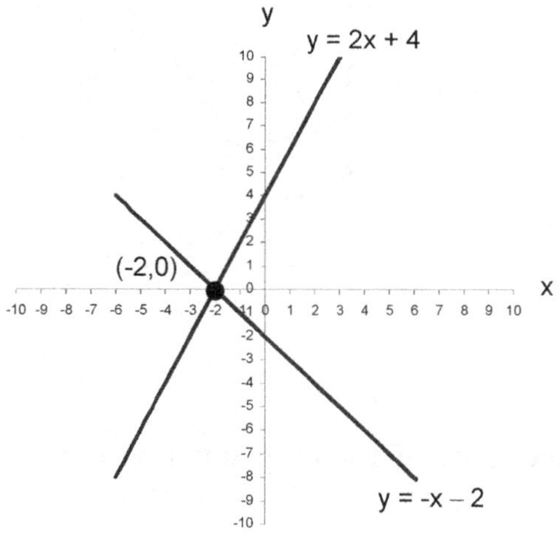

It should be noted that y = 0 also intersects each of these relationships at the point (-2,0).

Food for Thought
How can we know that 2 lines will only intersect at one point?

4.2.5 Puzzling it All Out

Learning to play a musical instrument, build a canoe, or any other worthwhile endeavor requires active involvement before the information provided can be fully integrated. The study of mathematics is no different.

Work each of these puzzles on your own until you are sure you have the correct answer. Then check the answer in the back of the book. If you are having trouble agreeing with the book's answer you can: check the solutions manual if one is available in your area or ask an online program to show you a solution, ask a friend or a tutor, or ask your teacher for advice. Keep in mind that there may be more than one way to solve each problem. Always look for the easiest solution.

It is preferable that you understand how to get the correct solution for every puzzle on this page before moving to the next section.

1. A line is parallel to $y=\frac{2}{3}x-6$ and has a y intercept at (0,4). Find its x intercept.

2. Give the slope and intercepts (if any) of each of the following equations:

    a. y = 4x - 2
    b. x = 3
    c. y = -5
    d. $d = rt + d_o$ (Slope and d intercept only)

3. Where do two perpendicular lines intersect if one can be represented by the equation $y=\frac{2}{3}x-5$ and the other has a y intercept at (0,8)?

4. What quadrant do the two functions in question 3 intersect? Imagine that the y axis in problem 3 represents money (in thousands of dollars) and the x axis represents time (in years). If one of the lines represents the depreciating value of my car over time and the other represents the balance on my car loan over time, what does the point of intersection tell you?

4.3 Chapter 4 - Review

In this chapter we looked at linear relationships in the 2D Cartesian coordinate system. We found that we could obtain a set of points representing an equation by substituting numbers for the independent variable (often x) and calculating the corresponding value of the dependent variable (often y). When these points are calculated, plotted, and connected for an equation involving multiplication by a single variable they form a line. For this reason we call these type of equations linear equations.

We found that we could represent any non-vertical line by the general formula

$$y = mx + b$$

represents a line in a 2D Cartesian Coordinate System

where x and y are any real numbers,

m = the slope of the line $\left(\dfrac{\Delta y}{\Delta x}\right)$,

and b = the y value of the y intercept of the line.

We looked at several lines and found that we could make some overall generalizations that facilitate interpretation including:

Lines from equations with positive slopes slope right,
Lines from equations with negative slopes slope left,
Horizontal lines have a zero slope, only a y intercept, and follow the form y = b,
Vertical lines have no slope, only an x intercept, and follow the form x = a

(where x any are any real numbers, b is the y value of the y intercept and a is the x value of the x intercept).

We looked at the special cases of parallel and perpendicular lines and found that

$m_1 = m_2$ Parallel Lines in Cartesian Coordinates

$m_1 = -\dfrac{1}{m_2}$ Perpendicular Lines in Cartesian Coordinates

where $m = \dfrac{\Delta y}{\Delta x}$ = the slope of the line.

We learned that when two lines intersect in Cartesian coordinates they share both x and y values at the point of intersection. This can only happen once for any two linear equations.

4.3.1 – More Practice Puzzles

1. Grace deposited a $100 gift in her savings account and now makes regular deposits of $15 per month. Her younger sister, Sally, opened her account at the same time with only $40 but has been regularly depositing $25 a month. Ignoring interest, how long will it be before Sally has more money in her account than Grace?

2. Write an equation for the graphed relationship. Label the slope and both intercepts.

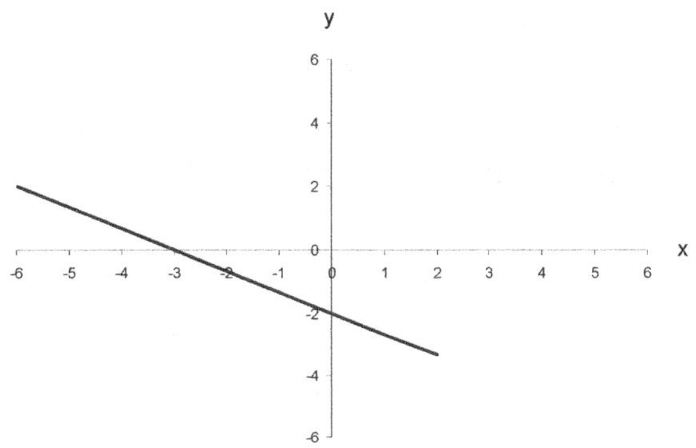

3. Graph the equation $y = 5\left(\frac{x}{9}+1\right)+x\left(\frac{7}{3}+\frac{1}{9}\right)-8+x$.

4. Find 3 equations that form parallel lines with y = -4x + 6.

5. Find 3 equations that are perpendicular to the equation in question 4.

6. Choose one of your equations from problem 4 and one from problem 5 and graph them. Find the point of intersection. Verify the point by making a chart of each equation. [Note: Scale intervals on each axis must be identical for the lines to appear perpendicular].

Chapter 5 – Quadratic Relationships in the Cartesian Coordinate System

As we know, the world is not just made up of straight lines and it would be nice to how to represent something that curves in the Cartesian coordinate system. To consider this, let us look at the simple linear relationship y = x. We can see a graph of this relationship yields a straight line through the origin with a slope of m = 1 as shown below.

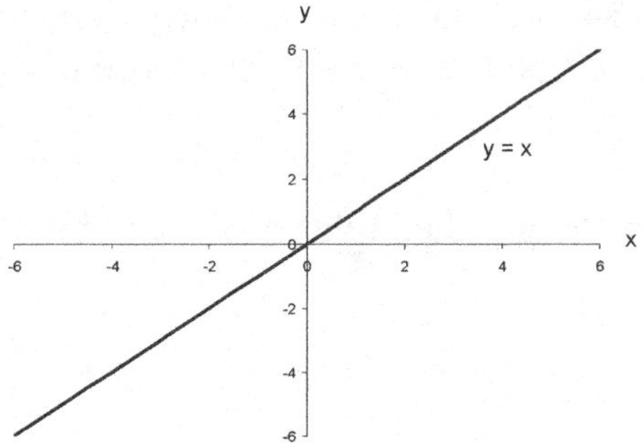

We know that adding any constant will just shift the line higher or lower on the graph:

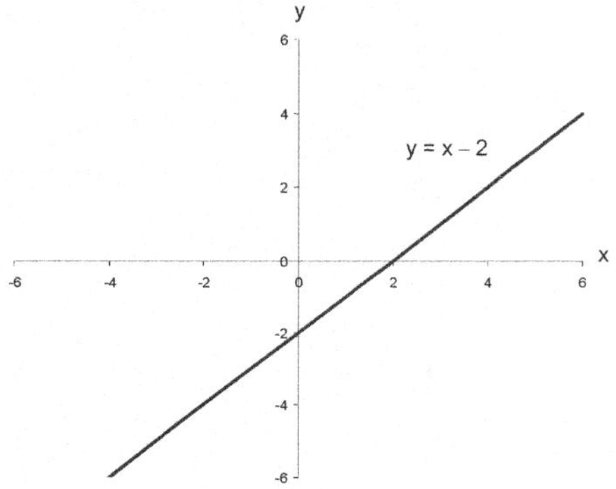

We also know that multiplying by a constant will just change the slope of the line:

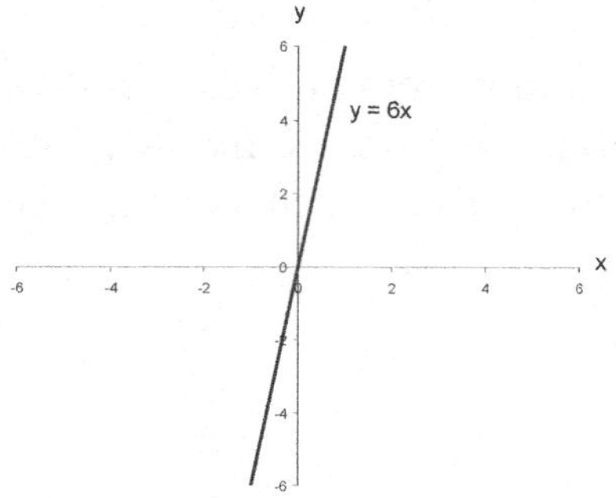

If we want something more than a uniform change, we must find a way to multiply the variable by more than just a constant. We have discovered that if we multiply the variable by itself, a curve is formed. For a variable that is multiplied by itself only once (x^2), we call the resulting form a parabola and the relationship quadratic.

5.1 – Charting a Parabola

We can make a chart and plot the points from a quadratic equation in the same fashion as we did for linear functions. A chart and corresponding graph for the simplest quadratic function, $y = x^2$ is shown below.

x	$y = x^2$
-4	16
-3	9
-2	4
-1	1
0	0
1	1
2	4
3	9
4	16

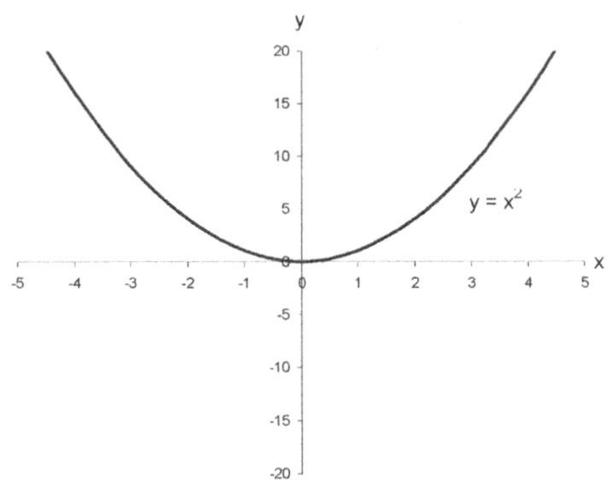

[Note: The x and y scales on our Cartesian graph have different intervals for clarity of presentation.]

Notice that the y values in the chart are identical for a given x value regardless of its sign (ie. $(-2)^2 = 2^2 = 4$ or $|2|^2 = 4$) giving the parabola 2 symmetrical sides.

The change is not uniform from one interval to the next (as illustrated below) resulting in curved as opposed to straight sides.

$x_2 - x_1$	$y_2 - y_1$
$1 - 0 =$ **1**	$1 - 0 =$ **1**
$2 - 1 =$ **1**	$4 - 1 =$ **3**
$3 - 2 =$ **1**	$9 - 4 =$ **5**

There is only one point that has a unique x and y value (0,0). This point is called the vertex of the parabola.

Food for Thought

What changes to the basic parabola equation shown above, $y = x^2$, might make shifts in its graph similar to those we have seen for linear equations?

5.1.1 Puzzling it All Out

Learning to play a musical instrument, build a canoe, or any other worthwhile endeavor requires active involvement before the information provided can be fully integrated. The study of mathematics is no different.

Work each of these puzzles on your own until you are sure you have the correct answer. Then check the answer in the back of the book. If you are having trouble agreeing with the book's answer you can: check the solutions manual if one is available in your area or ask an online program to show you a solution, ask a friend or a tutor, or ask your teacher for advice. Keep in mind that there may be more than one way to solve each problem. Always look for the easiest solution.

It is preferable that you understand how to get the correct solution for every puzzle on this page before moving to the next section.

1. Chart and graph the relationship $y = x^2 - 4$.
2. Chart and graph the relationship $y = -(x^2) + 4$.
3. What are the values of x when y = 0 for the relationship shown below?

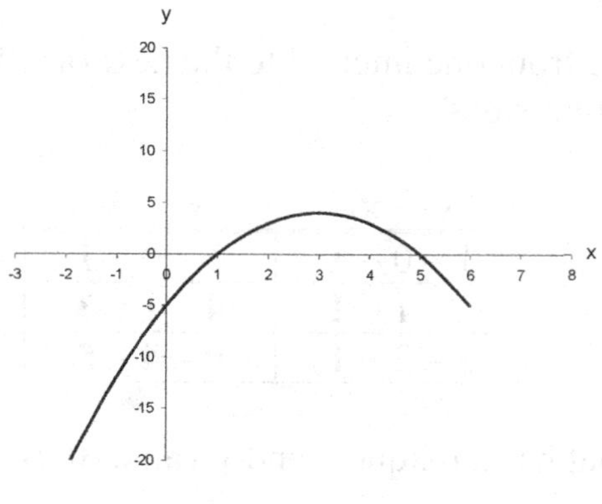

4. What is the value of the y intercept (x = 0) for the above relationship?

5.2 – Vertex – Curvature Formula for Parabolas

The puzzles in Section 6.1 revealed some interesting characteristics of quadratic equations in the Cartesian coordinate system (parabolas). A few of these are shown below.

A basic parabola with a positive squared variable forms a symmetrical arch about the y axis as we have seen.

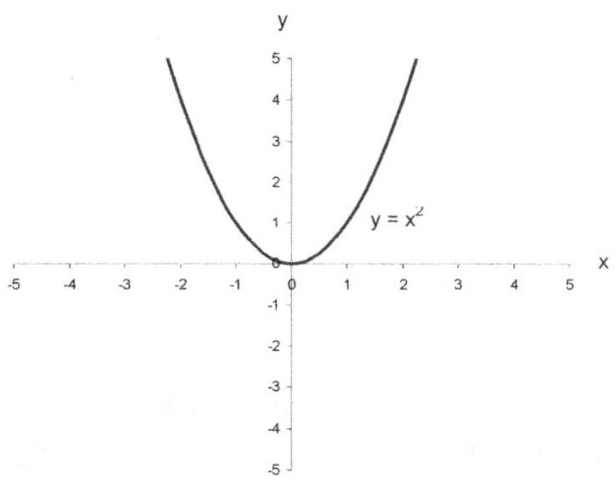

Adding a constant to a quadratic equation shifts the graph up or down on the y axis

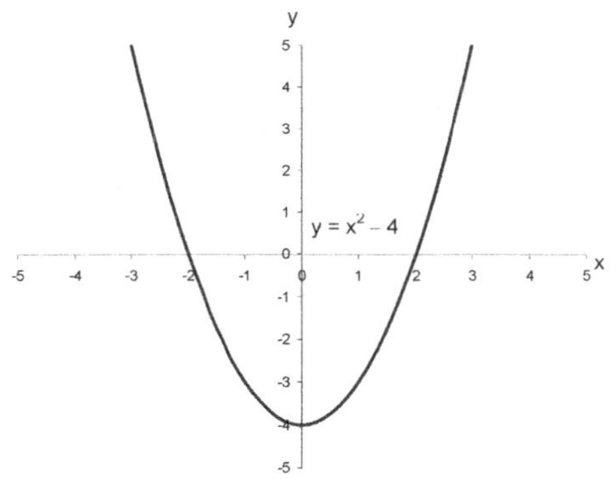

A quadratic equation with a negative squared variable will open the parabola downward.

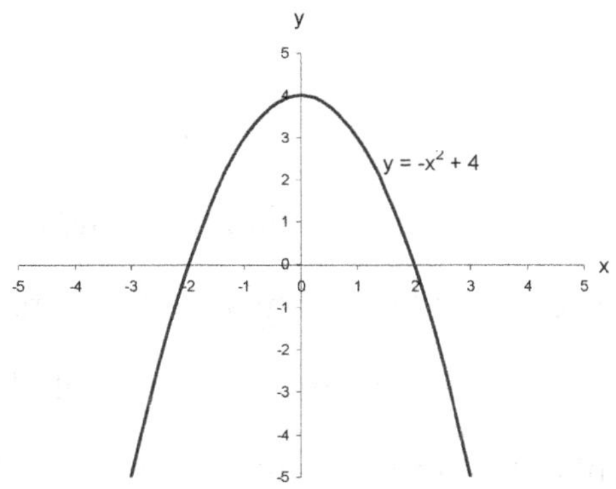

A further characteristic can easily be validated.

A coefficient to the squared variable, of other than 1, changes the curvature of the aperture (the width of the parabola opening)

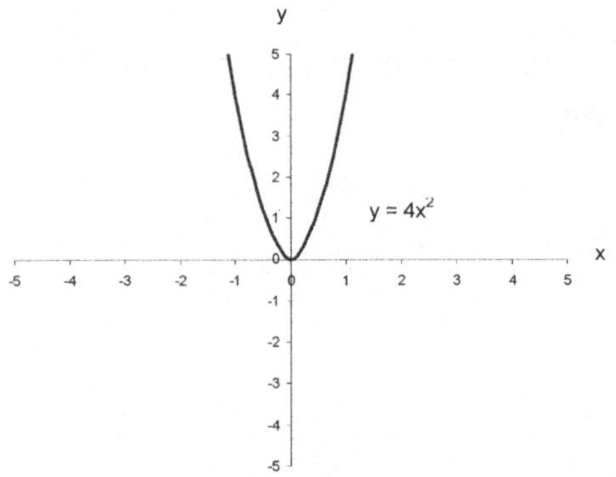

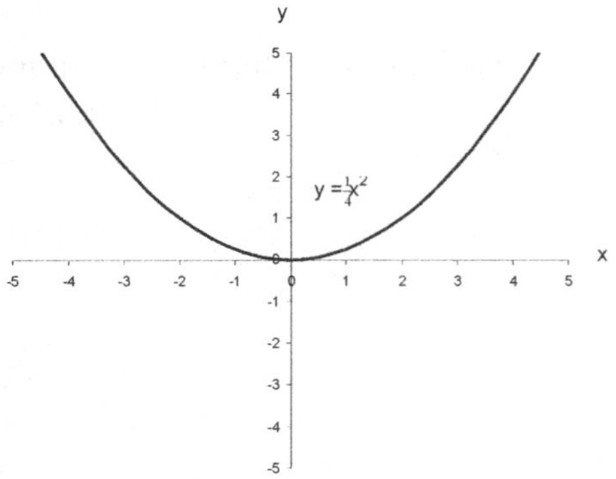

It is also possible to represent a parabola with a vertex on other than the y axis. We move the x input by adding a constant to the variable before squaring,

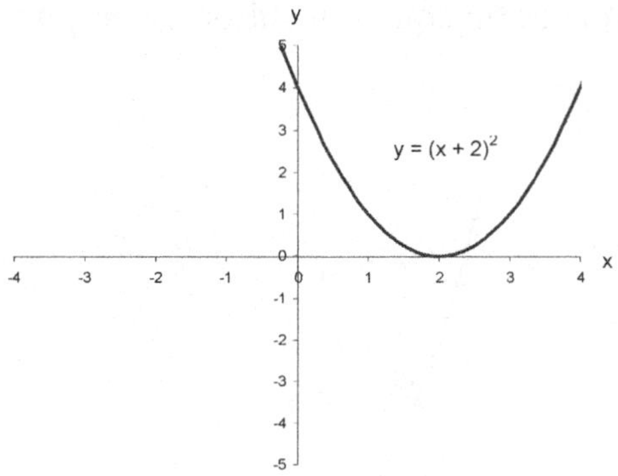

The understanding of these characteristics has led to the development of a standard general equation for quadratic relationships:

$$y = a(x-b)^2 + c$$

is the vertex-curvature formula for a standard parabola

where x and y are any real numbers,

the point (b,c) is the vertex of the parabola, and

a is a real number that modifies the curvature of the parabola.

5.2.1 Puzzling it All Out

Learning to play a musical instrument, build a canoe, or any other worthwhile endeavor requires active involvement before the information provided can be fully integrated. The study of mathematics is no different.

Work each of these puzzles on your own until you are sure you have the correct answer. Then check the answer in the back of the book. If you are having trouble agreeing with the book's answer you can: check the solutions manual if one is available in your area or ask an online program to show you a solution, ask a friend or a tutor, or ask your teacher for advice. Keep in mind that there may be more than one way to solve each problem. Always look for the easiest solution.

It is preferable that you understand how to get the correct solution for every puzzle on this page before moving to the next section.

1. Write an equation that describes the parabola below.

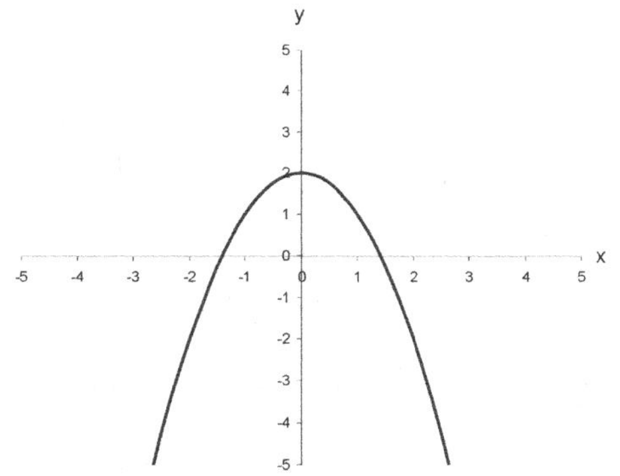

2. Sketch a rough drawing of the parabola represented by $y = -\frac{1}{10}x^2 + 3$.

3. Sketch a rough drawing of the parabola represented by $y = 10x^2 - 5$.

5.3 Chapter 5 – Review

We have seen that multiplying a variable by a constant produces a linear relationship (a straight line) in the Cartesian coordinate system while an exponential relationship (multiplying a variable by itself) results in a curve. The simplest of these relationships occurs when we multiply one variable by itself once ($y = x^2$) which results in the graph shown below:

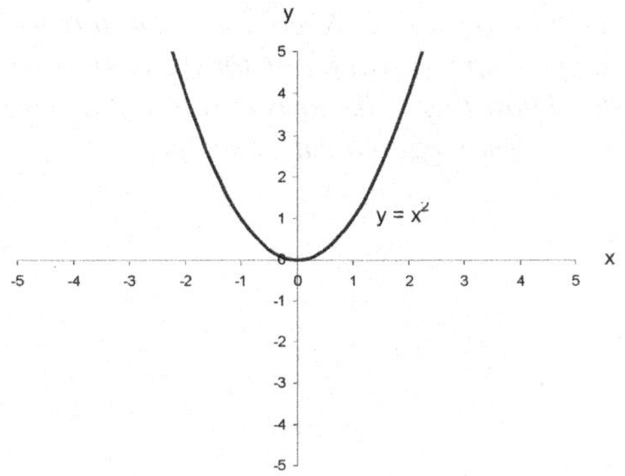

We looked at how changes to the equation affect its graph and developed the

Vertex-Curvature Formula for Parabolas,

$$y = a(x - b)^2 + c$$

where x and y are any real numbers,

the point (b,c) is the vertex of the parabola, and

a is a real number that modifies the curvature of the parabola.

Unit 3 – Finding Unknowns

We have already seen how we can use the formulas that we develop exploring relationships to find values that we did not previously know. The principles of equality discussed in Unit 3, along with the general principles reviewed in Unit 1, lead to a powerful tool in mathematics, the ability to rearrange an equation to solve for a unique unknown or make a formula more usable. The art of finding unknowns continues to be the most well known use for algebra.

Chapter 6 – Defining and Using the Properties of Equality

We know the meaning of an equal sign (=) from our arithmetic studies. When we say

$$5 + 3 = 8$$

we know that the expression on the left represents the same quantity as the expression on the right. We can also show easily through arithmetic that if we change one side of the expression, we must change the other side in a like manner for the equality to remain true. If we add or subtract to or from the left, we must add the same quantity to the right,

$$5 + 3 = 8 \text{ so,}$$

$$5 + 3 + 7 = 8 + 7 = 15 \text{ and}$$

$$5 + 3 - 4 = 8 - 4 = 4.$$

If we multiply or divide the left by some value we must do the same to the right,

$$2(5 + 3) = 2(8) = 10 + 6 = 16 \text{ and}$$

$$\frac{5+3}{2} = \frac{8}{2} = 4.$$

As might be expected, if we raise one side to a power, we must raise the other side to the same power in order to keep the equality true,

$$(5 + 3)^2 = 8^2 = 64.$$

As we learned in Unit 1, these properties are as applicable to variables as well as constants. As we will see, we can use this property to rearrange expressions giving us a powerful mathematical tool.

6.1 – General Statements about Mathematical Equality

Once we understand the property of equality, we can use the principles from Unit 1 to write several general mathematical statements that will assist us in rearranging formulas and solving for unknowns. The trick is to isolate the variable alone on one side of the equal sign. When the variable is alone, no further operations are performed on it and, if it is the only variable, we can easily condense what is left on the other side of the equal sign to a single number or simple rational form. If it is not a single variable, we will have gained a new formula that we can use. The mechanics are simple and the best part is that we have already learned them.

6.1.1 – Inverse Operations

First, we should remember that we defined subtraction, division, and roots as inverse operations of addition, multiplication, and exponents respectively. We made some general statements that illustrated what we meant by that definition such as:

Let a, b, and x = any real numbers

Then we can state that $a + b - b = a$
Inverse Operation of Addition (Subtraction),

$$\frac{ab}{b} = a$$

Inverse Operation of Multiplication (Division),

and $\sqrt[b]{\left(x^b\right)} = x$

Inverse Operation for Exponents (Roots).

Combining what we learn from these expressions with the principle of equality we discover our first basic tool. If we have an equation, we can isolate a variable from other operations by performing the inverse of the operation we want to negate. We can do this as long as we also perform the same inverse operation on the other side of the equality.

Example

Isolate the x in the equation 3 = 4x.

The x is being multiplied by a 4 so if we want to isolate it we must divide by 4. If we divide one side by 4 we must do the same to the other to maintain equality,

$$\frac{3}{4} = \frac{4x}{4}.$$

We can then simplify the expression and commute the equality to yield

$$x = \frac{3}{4}.$$

We may use the same technique with other operations provided that we use the proper inverse operation and remember to maintain equality.

6.1.2 – Inverse Order of Operations

Although the principles of inverse operations and equality give us a powerful tool, we are rarely confronted with equations with only one operation involved. We learned in Unit 1 that when more than one operation is involved in an expression the symbology of mathematics can present a problem. We learned that to resolve this problem we needed to perform higher order operations (those with embedded lower order operations) prior to those of lower orders and to perform any associated operations (those in parenthesis, etc) first. We presented the common anagram "PEMDAS", "Please Excuse My Dear Aunt Sally", to remind us of the correct order of operations needed to simplify or evaluate any expression.

This same issue with symbology arises when we try and use inverse operations to solve equations for we find that the order we follow in evaluating an expression is the exact opposite to the order we must use when disassembling it.

Example

Isolate the variable in the equation $2x + 4 = 10$.

To isolate the variable we have to perform the inverse operation to both sides of the equal sign, but there two different operations being performed. If we remember to use the inverse order of operations, the variable can be isolated in 2 steps.
First, we perform the inverse of the lowest order function,

$$2x + 4 - 4 = 10 - 4 \quad \text{gives}$$

$$2x = 6.$$

We can then divide both sides by 2 to isolate the x,

$$\frac{2x}{2} = \frac{6}{2} \quad \text{yielding} \quad x = 3.$$

The process is sometimes written showing vertical as opposed to horizontal operations as follows:

Example

Isolate the variable in the equation 2x + 4 = 10 using a vertical operation format.

The solution is in two steps as before but the inverse operations are written below the original statement and performed vertically as shown.

$$\begin{array}{rcl} 2x + 4 & = & 10 \\ -4 & & -4 \\ \hline 2x \quad & = & 6 \end{array}$$

We can then divide both sides by 2 to isolate the x as before,

$$\frac{2x}{2} = \frac{6}{2} \text{ yielding } x = 3.$$

Either form can be used although the vertical form is often easier to follow with complicated equations. Inverse operations may be applied through any number of operations in an equation provided that we remember to also inverse the order of operations.

6.1.3 – The Associative and Distributive Properties and Equations

When evaluating expressions we often used the Distributive Property to help us resolve expressions containing associated variables. There are several similar tricks that can help us address these quantities when they appear in equations.

Example

Isolate the variable y in the equation $y(4x-1)=12x-y$.

If we apply a strict reverse order of operations, we should first add the variable y to both sides to remove it from the left

$$y(4x-1)+y=12x-y+y$$
$$y(4x-1)+y=12x.$$

However, when we do so we are left with a complicated expression on the right that cannot be directly disassembled through inverse operations. We must first distribute the y across the associated quantity

$4xy-y+y=12x$ leaving

$$4xy=12x.$$

To isolate the y we must now divide both sides of the equation by 4x

$$\frac{4xy}{4x}=\frac{12x}{4x},$$

yielding y = 3.

A common mistake made by students is to forget that any expression that is a multiplicand or a divisor in a rational form is an implied association.

Example

Isolate the variable x in the equation $\dfrac{3x-5}{x+3}=5$.

Both expressions in the rational form are associated quantities with variables so we must treat them as though they were associated. This means that we cannot disassemble the lower order addition operations first. The solution to this problem becomes clear if we substitute a single variable for each of the associated quantities. If we let

$$y = 3x - 5 \quad \text{and} \quad f = x + 3,$$

the expression becomes $\frac{y}{f} = 5$.

We can then eliminate the rational form by multiplying both sides of the equation by f,

$$\frac{y}{f} f = 5f \quad \text{or} \quad y = 5f.$$

We can then substitute the original expressions back into the equation, simplify, and apply inverse operations as before.

$$3x - 5 = 5(x + 3)$$

$$\begin{array}{rcl} 3x - 5 &=& 5x + 15 \\ +5 && +5 \\ \hline 3x &=& 5x + 20 \\ -5x && -5x \\ \hline -2x &=& 20 \end{array}$$

$$\frac{-2x}{-2} = \frac{20}{-2} \quad \text{yields} \quad x = -10.$$

With practice, the substitution can be eliminated and the associated quantities can be moved directly:

$$\frac{3x - 5}{x + 3} = 5,$$

$$3x - 5 = 5(x + 3),$$

$$3x - 5 = 5x + 15,$$

$$3x - 5 + 5 = 5x + 20$$

$$3x - 5x = -2x = 20$$

$$\frac{-2x}{-2} = \frac{20}{-2} \quad \text{yields} \quad x = -10.$$

Formulas and expressions that we have developed or memorized can also help solve complicated equations. An example of this lies in the two expanded distributive expressions that we discussed in Unit 1,

$$(a+b)^2 = a^2 + 2ab + b^2 \qquad \text{and} \qquad (a+b)(a-b) = a^2 + b^2 .$$

If we recognize either of these forms in an equation, we can often substitute its equivalent to facilitate a solution.

Example

$$\textit{Solve } \frac{x^2 + 2x + 4}{x+2} = 6 \textit{ for x.}$$

In this case, even if we move the x+2 to the right, we will still be left with an x2 term that cannot be removed. However, if we recognize the perfect square in the multiplicand we can substitute yielding

$$\frac{(x+2)^2}{(x+2)} = 6 .$$

Using the rules for exponents we find we have

$$(x+2)^{(2-1)} = (x+2)^1 = x+2 = 6 .$$

Subtracting 2 from each side isolates the variable and gives us x = 4.

In Chapter 9 we will learn methods that will help us resolve many complex associated quantities without memorizing. However, remembering the formulas for the difference of squares, perfect squares, or similar common forms can often speed up problem solution.

6.1.4 Puzzling it All Out

Learning to play a musical instrument, build a canoe, or any other worthwhile endeavor requires active involvement before the information provided can be fully integrated. The study of mathematics is no different.

Work each of these puzzles on your own until you are sure you have the correct answer. Then check the answer in the back of the book. If you are having trouble agreeing with the book's answer you can: check the solutions manual if one is available in your area or ask an online program to show you a solution, ask a friend or a tutor, or ask your teacher for advice. Keep in mind that there may be more than one way to solve each problem. Always look for the easiest solution.

It is preferable that you understand how to get the correct solution for every puzzle on this page before moving to the next section.

1. Isolate the variable x in the equation $\frac{2}{3}x = 4$.

2. Solve y + x = 2 for the variable y.

3. Isolate the variable x in $\frac{2}{3}x + 2 = 6$.

4. Isolate the variable r in $d = rt + d_0$.

5. Solve for x in $\frac{x^2 - 9}{x - 3} = 2x - 5$.

6. Solve for y in $2y(y + 4) = y^2 + 8y + 16$.

6.2 – Rearranging Formulas

One use of the properties of equality is to rearrange formulas that have been developed. For instance, we may find a relationship between the cost of bananas and the distance that they are shipped to be $B = B_o + rd$ where B is the shipped cost of the bananas, B_o is the banana cost before shipping, d is the distance to be shipped, and r is the shipping rate. This is a useful formula if one is routinely asked for the shipped cost of bananas for a given distance. However, if one is often asked instead how far the bananas can be shipped for a given overall cost, a rearranging of the equation provides a direct solution that can be used repetitively,

$$d = \frac{B - B_o}{r}.$$

[Note: The student is encouraged to verify that this is, indeed, still an equality based on the original.]

6.2.1 – Geometric, Scientific, Business, and other Formulas

We can rearrange any formula that we have developed to repeatedly solve for any variable in the formula. The process is simple and we will do it here for a few of the formulas that we developed in Unit 1.

Example

You are head of the community garden club and this year there are 6 different widths available for beds. Members can have any length they want but must pay a charge for beds that are over 100 feet in total distance around (Perimeter). You would like to make a chart that shows the maximum length for each width that can be had without a fee. You have the equation you developed in your Algebra Class

$$P = 2(L + W)$$

(where L, W, and P are the length, width, and Perimeter of a rectangle respectively).

You would prefer not to have to solve this equation for L for each of the 6 widths. Instead, you want to rearrange the equation to isolate L and use the new formula directly.

We first notice that there is an associated quantity containing variables. It makes no sense to move it as a unit so we will use the distributive property to remove the association,

$$P = 2(L + W) = 2L + 2W.$$

Inversing lower order operations first we move the 2W to the left side of the equation,

$$P - 2W = 2L.$$

We can then divide by 2 to isolate the L. Remember, we must divide everything on each side of the equation by 2 to maintain equality,

$$L = \frac{P - 2W}{2}.$$

Since we know that P = 100 ft is the threshold we are looking for in our chart, we can further tailor the equation to our needs,

$$L = \frac{100 - 2W}{2}$$

(where L and W, are the length and width, respectively, of the garden space in feet).

We can now input the various widths into the equation and calculate each quickly and directly.

Perhaps we would like an equation that can quickly tell us what size radius we must use to draw a circle with a particular area. We recall from Unit 1 that the area of a circle can be expressed in terms of the radius as $A = \pi r^2$. A few steps, starting with the lower order first, can provide an equation for the radius in terms of the area as shown below.

$$A = \pi r^2$$

$$\frac{A}{\pi} = \frac{\pi r^2}{\pi} \text{ gives } \frac{A}{\pi} = r^2$$

$$\sqrt{r^2} = \sqrt{\frac{A}{\pi}}$$

$$r = \sqrt{\frac{A}{\pi}}.$$

We developed the common formula for temperature conversion from degrees Celsius to degrees Fahrenheit. Although this formula is quite useful for people routinely using the English Unit system, those using the metric system would be better served by rearranging

$$F = \frac{9}{5}C + 32$$

$$F - 32 = \frac{9}{5}C$$

$$\frac{5}{9}(F-32) = C \quad \text{or} \quad C = \frac{5}{9}(F-32).$$

6.2.2 – Mathematical Formulas (Including Point-Slope Formula for a Line)

Mathematical formulas describing relationships in the Cartesian coordinate system can also be rearranged once they are derived. We can also rearrange existing formulas into standard formats so that they may be easily graphed or solved.

Example

Graph the relationship $\frac{6}{y} = \frac{2x-2}{3x^2-6x+3}$.

Although the relationship looks like it might be linear, it needs a lot of rearranging before we can graph it. We first apply the distributive principle, in reverse, to obtain

$$\frac{6}{y} = \frac{2(x-1)}{3(x^2-2x+1)}.$$

If we recognize that $x^2 - 2x + 1 = (x-1)^2$, we can make the substitution

$$\frac{6}{y} = \frac{2(x-1)}{3(x-1)^2}.$$

We can then use the rules of exponents to obtain

$$\frac{6}{y} = \frac{2}{3}(x-1)^{(1-2)} = \frac{2}{3}(x-1)^{(-1)} = \frac{2}{3(x-1)} \quad \text{or} \quad \frac{6}{y} = \frac{2}{3(x-1)}$$

We need to bring the variables to the top of the rational form. Since they are currently being divided we can achieve this through multiplication

$$\frac{6}{y}y = \frac{2}{3(x-1)}y \quad \text{gives} \quad 6 = \frac{2y}{3(x-1)}$$

and then $\quad 6(3)(x-1) = \frac{2y}{3(x-1)}3(x-1)$ yielding

$$18(x-1) = 2y.$$

Distributing the 18 throughout the associated quantity we get

$$18x - 18 = 2y.$$

We can then divide both sides of the equation by 2 to isolate the y

$$y = \frac{18x-18}{2} = 9x - 9.$$

Once we have the equation in slope intercept form we can readily graph it using the standard general form to find the slope and y intercept

y = mx + b

↓ ↓

y = 9x – 9

Since we know we have the y intercept (0,-9) we can plot this point. We can then use the slope to obtain a second point. The equation $m = \frac{\Delta y}{\Delta x} = \frac{9}{1}$ tells us that for every 9 intervals we go up we must go 1 interval to the right to get back to the line. Since the y intercept is exactly 9 units below the origin, a second point, (1,0) is readily available and we can graph the equation as shown below.

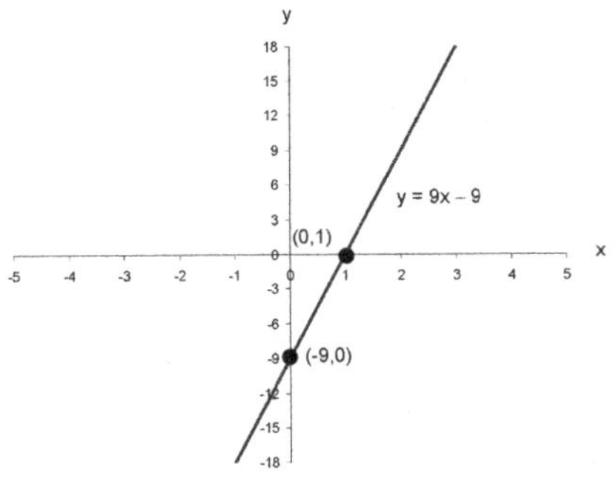

Once we have the graph we can use it to find specific points or to describe trends in the relationship to others.

Point-Slope Formula of a Line

The slope intercept form of a linear equation is a powerful tool for graphing, but occasionally we need other forms. The point slope formula for a line is often derived from the formula for the slope follows:

If we look at the graph we developed in the prior example we will notice that the slope can be defined in terms of the x and y values of the points we have chosen to graph

$$m = \frac{\Delta y}{\Delta x} = \frac{(9-0)}{(1-0)} = 9 .$$

A closer look will reveal that this is true for any two points on the line. We can, therefore, rewrite the formula for the slope in terms of any two points, (x,y) and (x_1,y_1), where (x_1,y_1) is a known point on the line and (x,y) represents any unknown point on the line as

$$m = \frac{\Delta y}{\Delta x} = \frac{(y-y_1)}{(x-x_1)} .$$

Since the equation contains both the slope and at least one point, it fully defines the line. The equation is generally expressed as

$$m(x-x_1) = (y-y_1)$$

Point – Slope Formula for a Line
(where m = the slope and (x_1,y_1) and (x_2,y_2) are points on the line)

and is often used to obtain the equation for a line when the y intercept is unknown but two other points are readily available.

6.2.3 Puzzling it All Out

Learning to play a musical instrument, build a canoe, or any other worthwhile endeavor requires active involvement before the information provided can be fully integrated. The study of mathematics is no different.

Work each of these puzzles on your own until you are sure you have the correct answer. Then check the answer in the back of the book. If you are having trouble agreeing with the book's answer you can: check the solutions manual if one is available in your area or ask an online program to show you a solution, ask a friend or a tutor, or ask your teacher for advice. Keep in mind that there may be more than one way to solve each problem. Always look for the easiest solution.

It is preferable that you understand how to get the correct solution for every puzzle on this page before moving to the next section.

1. Using the formula derived in Unit 1 for the area of a triangle, find an expression that will return the length of the base for a triangle with a given height and Area.

2. Using the formula A = P(1 + rt) where A= the total amount we have at the end of the investment, P = principle (amount of money invested), r = rate, and t = time, develop an expression that will give the rate of interest required to return a particular amount on an investment (hint - isolate the r).

3. Graph the relationship $x(7x+3) = x(y+7x)$..

4. Write an equation for a line that goes through the 2 points, (-3, -4) and (2,6). What is the value of this equation when y = 0 (at the y-intercept)?

6.3 Chapter 6 – Review

In Chapter 6 we have introduced the basic properties of equality and inverse operations and developed methods that allow us to rearrange equations and solve for unknowns. We observed that the equality of an equation will remain true provided that any changes made on one side of the equation are made in the same manner on the other.

We noted that equality combined with the principles of inverse operations:

$$a + b - b = a$$

Inverse Operation of Addition (Subtraction),

$$\frac{ab}{b} = a$$

Inverse Operation of Multiplication (Division),

and $\sqrt[b]{\left(x^b\right)} = x$

Inverse Operation for Exponents (Roots)

(where a, b, and x are any real numbers),

allow us to rearrange equations to suit our needs provided that we perform each inverse operation in the reverse of the Order of Operations we discovered for evaluating expressions. Therefore we must undo lower order operations first saving associated quantities (parentheses) for last.

We also looked at using the Distributive Property both backward and forward to help us simplify expressions before rearranging. Recognizing forms such as perfect square trinomials (expressions with 3 terms) or the difference of squares (binomials) can further help us simplify expressions making equations easier to rearrange and solve.

In Section 2 we used our new tools to develop many new scientific, geometric, and business formulas by rearranging those we derived in Unit 1. We also looked at how mathematical formulas can be rearranged to assist with graphing and developed the Point-Slope Formula for a line

$$m(x - x_1) = (y - y_1)$$

Point – Slope Formula for a Line

(where m = the slope and (x_1,y_1) and is a point on the line).

6.3.1 – More Practice Puzzles

1. Solve $\frac{3x^2-27}{4x+12}=6$ for x.

2. Liquid will rise due to surface tension in a tube to a height of

$$h=\frac{2\gamma}{\rho g r}$$

where h is the height of the liquid in the tube, γ is the surface tension of the liquid, ρ is the density of the liquid, g is the acceleration of gravity, and r is the tube radius.

Rearrange the formula to an equation that can be used to calculate the tube radius required to raise a liquid to specific heights.

What do you notice about the relationship between the radius and the height? Put some constants into the equation to validate your answer.

3. Graph the relationship

$$\frac{3xy-3y}{(\sqrt{x}+1)(\sqrt{x}-1)}-5=\frac{y}{2}.$$

Chapter 7 – Solving Linear Equations

We have defined a linear equation as one that forms a line when plotted in the Cartesian coordinate system. Multiplication of a variable by a constant produces a uniform change. Equations that follow the formula $y = mx + b$ produce such a relationship. Although linear equations may take many forms, two variable equations can always be reduced to the $y = mx + b$ format and will always involve only variables with an exponent of 1 (or 0).

7.1 – Solutions in Single Variable Linear Relationships

If there is only one variable with an exponent of 1 in the equation, the equation can be rearranged to obtain a constant solution. We have done this several times in Chapter 7 and the methods are the same. In this section we will look at how the properties of equality and operations provide us with other useful tools for solving these equations.

7.1.1 – Simple Linear Equations

First, we will review the rearranging of a simple linear equation with one variable to provide a constant solution.

Example

Solve 4x – 2 = 7 for x.

We begin with the lowest order operation, in this case subtraction. Since addition is the inverse of subtraction we will add 2 to each side to isolate the 4x as shown below.

$$\begin{array}{rcl} 4x - 2 & = & 7 \\ +2 & & +2 \\ \hline 4x & = & 9 \end{array}$$

We can then divide both sides by 4 to isolate the x

$$\frac{4x}{4} = \frac{9}{4} \quad \text{yields} \quad x = \frac{9}{4}.$$

[Note: The answer is left in rational form with the lowest possible terms.]

If we look closely at the equation 4x – 2 = 7, we will see that it describes a linear relationship in y = mx + b format (y = 4x – 2) at a specific point where y = 7. If we graph each of these equations, as done below, we can quickly verify that the x value of the point of intersection is indeed just slightly over 2 at $(\frac{9}{4}, 7)$.

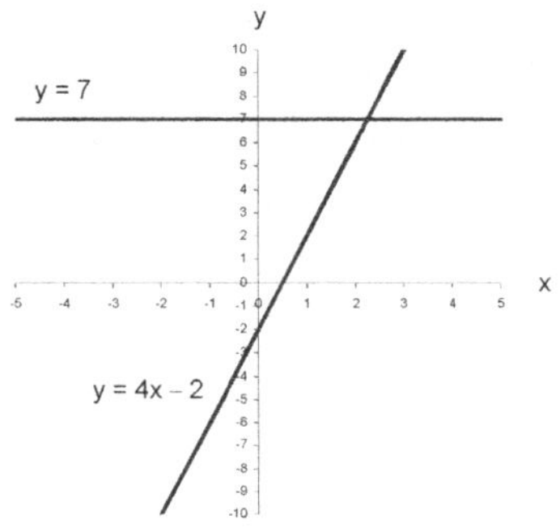

7.1.2 – Equations with Rational Expressions or Decimal Constants

We have seen many equations with rational forms. In each case we have tried to isolate or simplify the rational form and then multiply both sides by the divisor to remove them. In reality, the property of equality provides us with a much simpler method of dealing with these more complicated linear equations.

Example

$$\textit{Solve } \frac{2}{3}x + \frac{1}{6} = \frac{2}{9} \textit{ for x.}$$

To solve this equation as we have done in the last Chapter, we would subtract $\frac{1}{6}$ from each side and then multiply the answer by 3 and divide by 2. This will involve converting both fractions to a common divisor before subtracting. The process is tedious, time consuming, and highly subject to err. Instead, we will use the property of equality (equality is maintained if we perform like operations on each side) to remove the rational forms directly.

We know that we can get rid of the divisor in each term by multiplying the term by it;

$$\frac{2}{3}x(3) = 2x,\ \frac{1}{6}(6) = 1,\ \frac{2}{9}(9) = 2.$$

This doesn't help because the terms must be multiplied by different constants and we must have the same to maintain equality. We might note, however, that while $\frac{2}{3}x(3) = 2x$

removes the rational form, $\frac{2}{3}x(6)=4x$ does also. In fact, multiplying by any even multiple of 3 will remove the rational form of $\frac{2}{3}$.

If we factor each divisor into its prime factors (find the prime numbers that, when multiplied, will result in the divisor), we can find a number that we can multiply by each side of the equation to remove the rational forms. Since we want the lowest possible number, we use only the product of the independent prime factors for each divisor.

[Note: A product that includes all prime factors would also work but is unnecessary.]

$$\frac{2}{3}x+\frac{1}{6}=\frac{2}{9}$$

Divisor	Prime Factors	Independent Prime Factors
3	3, 1	3, 1
6	3, 2, 1	2
9	3, 3 , 1	3

After factoring we find that multiplying each side of the equation by 3(2)(3)(1) = 18 should clear all rational forms. Performing this operation we get

$$18\left(\frac{2}{3}x+\frac{1}{6}\right)=\left(\frac{2}{9}\right)18.$$

Applying the distributive property and multiplying yields:

$$\frac{2x}{3}(18)+\frac{1}{6}(18)=\left(\frac{2}{9}\right)18 \quad \text{and} \quad \text{gives} \quad 2x(6)+3=4.$$

Solving:

$$12x+3=4$$

$$12x=1$$

$$x=\frac{1}{12}.$$

Note that the method will work regardless of whether a constant or variable is the divisor.

Example

Solve $\frac{4}{x}+\frac{2}{3}=\frac{5}{6}$ *for x.*

We first factor the each divisor to find a useful multiplier.

Divisor	Prime Factors	Independent Factors
x	x, 1	x, 1
3	3, 1	3
6	3, 2 , 1	2

We then multiply both sides of the equation by x(3)(2) = 6x to clear the rational forms.

$\frac{4}{x}(6x)+\frac{2}{3}(6x)=\frac{5}{6}(6x)$ and get $24+4x=5x$.

Solving, x = 24.

A similar procedure can be used when solving equations with decimals. Multiplication of each side of the equation by an appropriate multiple of 10 eliminates the decimals as shown below.

Example

Solve $0.15x+2.5=3.25$ *for x.*

$100(0.15x+2.5)=3.25(100)$ yields 15x + 250 = 325.

Subtracting 250 leaves 15x = 75 or x = 5.

7.1.3 – Using Linear Equations to Solve Problems

To make equation solving useful, we must often find ways to translate real life problems into solvable equations. The first step in this endeavor is to assign, or define, a variable to represent the unknown quantity we are trying to find. The other information we know can then translated into a mathematical statement that represents the relationship and the equation rearranged, if necessary, to find the value of the unknown. An example is given below.

Example

Alex and Joel wish to box pears from their fruit trees for gifts. They want to put together 8 boxes in total and save 3 pears to eat with a friend after they make the boxes. If they have 27 pears, how many should they put in each box?

Although you can easily guess the answer to this problem, it provides us with a good example of the method. First, we read through the information and decide what it is we do not know that we wish to know. The answer here is provided in the last line of the question. How many pears go each box. We assign a variable to this quantity:

p = the number of pears in each box

Once we have assigned a variable we can build a mathematical statement about it. We know that the total number of pears will be the amount in the boxes (8p) plus the pears that they want to reserve to eat later (-3). We also know that this same total equals 27. Setting the two equal to each other,

$$8p - 3 = 27.$$

The resulting equation can be easily rearranged to isolate the p,

$$\begin{array}{r} 8p - 3 = 27 \\ +3 \quad +3 \\ \hline 8p \quad = 24 \end{array}$$

$$\frac{8p}{8} = \frac{24}{8} \qquad \text{yields} \qquad p = 3.$$

They should put 3 pears in each box.

Although most problems that result in single variable linear solutions can often be solved through logic faster than formal algebraic methods, they provide simple, easy to verify, practice. Once mastered, the method can be applied to many complex problems that would be cumbersome (if not impossible) to resolve without algebra. Students are encouraged to practice writing algebraic equations to solve problems as often as possible until they are very comfortable with the method.

7.1.4 Puzzling it All Out

Learning to play a musical instrument, build a canoe, or any other worthwhile endeavor requires active involvement before the information provided can be fully integrated. The study of mathematics is no different.

Work each of these puzzles on your own until you are sure you have the correct answer. Then check the answer in the back of the book. If you are having trouble agreeing with the book's answer you can: check the solutions manual if one is available in your area or ask an online program to show you a solution, ask a friend or a tutor, or ask your teacher for advice. Keep in mind that there may be more than one way to solve each problem. Always look for the easiest solution.

It is preferable that you understand how to get the correct solution for every puzzle on this page before moving to the next section.

1. Solve $\frac{2}{3}x-4=6$ for x.

2. Solve $\frac{2q}{7}-\frac{3}{14}=\frac{3}{2}$ for q.

3. You would like to make 12 identical geometric shapes with equal length sides. You have 325 ft of balsa wood for the construction but 1 ft of it has an undesirable coloration. How long should each side be to make full use of the remaining wood?

4. Christine wants to make wall hangings for a local festival. Each hanging must be cut from $\frac{5}{6}$ of a full sheet of material. There is border of 1" ribbon $\left(\frac{1}{12}ft\right)$ along both the top and the bottom of each hanging. If each hanging must carefully fit into alcoves that are $3\frac{1}{2}$ ft long, how long do the full sheets of material have to be?

7.2 – Solutions in Multi Variable Linear Relationships

We saw in Unit 2 that when two linear relationships are graphed in a Cartesian coordinate system they will always intersect at exactly one point unless the lines are parallel. At this point, both equations yield the same y value for a particular x. In Unit 2 we found this point graphically, but, as with single variable equations, it is also possible to solve the equations algebraically. To find the common solution point, both equations are required. There are two generally accepted algebraic methods for multi-variable linear equation solving, the substitution method and the elimination method.

7.2.1 – Substitution Method

When at least one of the equations is in terms of one of the variables, it is convenient to substitute the function for the variable into the second equation and solve.

Example

Find the common solution point for

$y = 2x + 3 \qquad \textit{and} \qquad 7 = 4x + 3y.$

Since we know that $y = 2x + 3$, we can substitute 2x + 3 for y in the other equation.

$$7 = 4x + 3(2x + 3).$$

With only one variable in the equation we can simplify and solve.

$$7 = 4x + 3(2x + 3) = 4x + 6x + 9 = 10x + 9$$

$$\begin{array}{rcl} 7 & = & 10x + 9 \\ -9 & & -9 \\ \hline -2 & = & 10x \end{array}$$

$$\frac{-2}{10} = \frac{10x}{10} \qquad \text{yields} \qquad x = -\frac{1}{5}.$$

Once we know the x value we can use either original equation to obtain the y value and the other to verify.

$$y = 2\left(-\frac{1}{5}\right) + 3 = \frac{-2}{5} + \frac{15}{5} = \frac{13}{5},$$

$$7 = 4\left(-\frac{1}{5}\right) + 3\left(\frac{13}{5}\right) = \frac{-4}{5} + \frac{39}{5} = \frac{35}{5} = 7 \checkmark.$$

7.2.2 – Elimination Method

A method that eliminates the need to isolate one of the variables for substitution is known as the elimination method. This method is grounded in the properties of equality. To review slightly, the symbol, =, in mathematics tells us that the expressions appearing on either side of the symbol are the same, regardless of their form. We can substitute the expression on one side of the equal sign for the value of the other as we have done in the substitution method. We also know that we may perform any operation on one of the expressions and the equality will remain true provided we perform the identical operation on the other.

The equation, 2e + 2f = 8 expresses a linear relationship between two variables. We know tha we could add 4 to each side of this equation and the equality would remain true. If we have another equation, 3e – 2f = 4, we can readily add the constant 4 directly to the right side of the equation (to the 8). In addition, we note that adding 3e – 2f (a substitute form of 4) to the left side of the equation results in the elimination of one of the variables as shown below:

$$\begin{array}{rcl} 2e + 2f & = & 8 \\ +4 & & +4 \\ \hline \end{array}$$

$$\begin{array}{rcl} 2e + 2f & = & 8 \\ +\underline{(3e - 2f)} & & \underline{+4} \\ 5e & = & 12 \end{array}$$

The value for e may then be found and back substituted into either original equation to find f as shown in the substitution example.

The last step shown above is clearly just an addition of the two equations. It can be readily shown (and the student is encouraged to investigate) that any two linear equations may be added, subtracted, multiplied or divided by one another as required to eliminate variables and still maintain equality.

This method can be extended to linear equations with any number of variables, known as a system of equations. Each variable is successively removed through a combination of equations until the value for a single variable can be obtained. Values are then successively back substituted until each has been found. For a complete solution, a separate and independent equation must be provided for each variable to be solved.

Computers are often used today to perform this analysis on problems with thousands of unknowns. The method has allowed us to perform in depth analysis on things like heat and stress transfer, the source of sounds in space, and the nature of an atom.

Food for Thought

Why must there be one independent equation for every variable in the solution?

Example

Solve the system of equations $6x - 3y = -9$ *and* $2x + \frac{3}{2}y = \frac{7}{2}$.

The second equation would be easier to work with without the rational forms. Multiplying each side of the equation by 2 yields a simpler expression,

$$2\left(2x + \frac{3}{2}y\right) = \left(\frac{7}{2}\right)2$$

$$4x + 3y = 7.$$

Since one equation has a term with 3y and the other with -3y, we can add one equation to the other and solve:

$$\begin{array}{r} 4x + 3y = 7 \\ +\ (6x - 3y) = -9\ . \\ \hline 10x \qquad = -2 \end{array}$$

$$x = -\frac{1}{5}\ .$$

Back substitution requires isolating the variable,

$$4\left(-\frac{1}{5}\right) + 3y = 7$$

$$5\left[\left(-\frac{4}{5}\right) + 3y\right] = (7)5$$

$$-4 + 15y = 35$$

$$15y = 39$$

$$y = \frac{39}{15} = \frac{13}{5}.$$

We can validate our answer with the other original equation,

$$2\left(-\frac{1}{5}\right) + \frac{3}{2}\left(\frac{13}{5}\right) = \frac{7}{2}$$

$$10\left(2\left(-\frac{1}{5}\right) + \frac{3}{2}\left(\frac{13}{5}\right)\right) = \left(\frac{7}{2}\right)10 = -4 + 39 = 35. \checkmark$$

7.2.3 Puzzling it All Out

Learning to play a musical instrument, build a canoe, or any other worthwhile endeavor requires active involvement before the information provided can be fully integrated. The study of mathematics is no different.

Work each of these puzzles on your own until you are sure you have the correct answer. Then check the answer in the back of the book. If you are having trouble agreeing with the book's answer you can: check the solutions manual if one is available in your area or ask an online program to show you a solution, ask a friend or a tutor, or ask your teacher for advice. Keep in mind that there may be more than one way to solve each problem. Always look for the easiest solution.

It is preferable that you understand how to get the correct solution for every puzzle on this page before moving to the next section.

1. The 8 o'clock am train for Boston left before the track was realigned and is heading at a rate of 70 mph on the wrong track. The train to Worchester left on the track the Boston train is now traveling on at 7:30 am and is traveling at only 50 mph. Assuming both trains can stop on a dime, how long do the station masters have to warn the conductors on the trains of an impending collision? How far will each train have traveled when they collide?

   [Hint – Use the formula $d = rt + d_0$ to write 2 equations, then solve for t]

2. Solve the set of equations for p and q:

   $$p - 1 = q \qquad \text{and} \qquad -2p + 8 = q\,.$$

7.3 Chapter 7 – Review

Chapter 7 gave us a chance to look at methods to solve linear equations, equations that can be put into the form y = mx + b.

We saw that linear equations with only variable, such as

$$4x - 6 = 2 \text{ or}$$

$$\frac{3x-5}{12}+\frac{x^2-25}{6(x+5)}=9(x-4),$$

will always resolve to a single solution.

We looked at ways to rearrange equations with rational forms and decimals by multiplying each side of the equation by a term that will remove these from the entire equation.

We looked at a process for writing equations based on relationships in problems we were trying to solve. We found that in most cases, it is best to assign a variable to the quantity that we are looking for and construct the equation based on the relationship of other values in the problem to the variable.

We saw that with at least one equation per variable we could use the properties of equality to systematically

eliminate variables,
solve for the last value, and
back substitute each value

until the entire set of equations is solved. We looked at two methods to solve simultaneous equations (equations representing the same variables) and used them to solve sets of linear equations with two variables.

The substitution method involves

solving one equation for one of the variables and
substituting the expression for it into the second equation.

The elimination method utilizes

addition, subtraction, multiplication, or division of one of the equations to the other

with elimination of one of the variables as a goal. The elimination method has been adapted to computer programs for the solution of equation sets with thousands of variables.

7.3.1 – More Practice Puzzles

1. Solve $24.8x = 0.62y + 1.24$ for x.

2. Solve the equation 7(r – 2) – 2(3 + r) = 2 – r.

3. Solve the equation $\frac{1}{3} + \frac{x}{2} + \frac{2}{3}y = 5$ for y.

4. Solve the equation $\frac{x^2 - 8x + 16}{x - 4} = 0$ for x.

5. Where does the line with a slope of -4 and an included point of (0,3) intersect the line that runs between the points (-3,0) and (2,5)?

6. A boy finds 3 envelopes with an equal amount of cash in each in a bag with $125 in small bills. After he turns it in, the total amount of money reported was $800. Write an equation to represent this situation and solve it to find how much money was in each envelope.

Chapter 8 – Solving Quadratic Equations

As discussed in Unit 2, quadratic equations are equations that have a variable with an exponent of 2. They form a parabola in the Cartesian coordinate system and can be expressed in the

> vertex-curvature format for a quadratic equation,
> $$y = a(x-b)^2 + c$$
> where x and y are any real numbers,
> the point (b,c) is the vertex of the parabola, and
> a is a real number that modifies the curvature of the parabola.

Unlike a linear relationship that always has one unique value of x for each value of y; quadratic equations have regions where each value of x can return none, one, or two values for y. A plot of the simplest quadratic equation, $y = x^2$, is shown below. In the striped area (and above) where the parabola exists above the vertex, it can be seen that two x values will result in the same y value. There is only one unique value for x and y at the vertex (the line y = 0) and below the vertex there is no parabola so there are no y values in this region that fall on the parabola regardless of the x value.

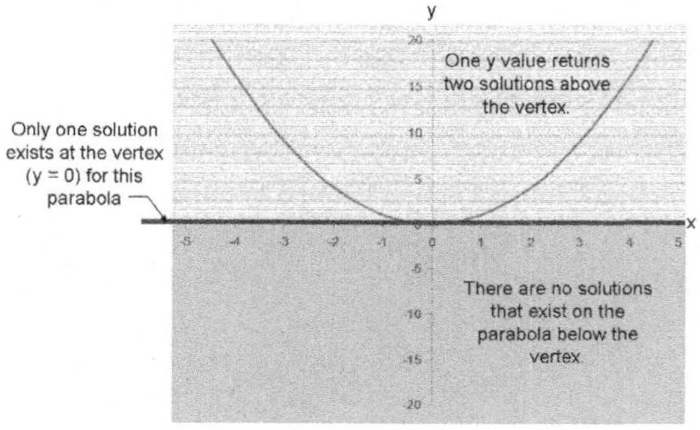

Quadratic equations with one variable describe a quadratic relationship at a single y value. These types of equations may have one, two, or no solutions depending on the particular parabola and y value they represent.

8.1 – Square Root Method

The simplest form of quadratic equation also gives us the best place to start looking for solutions to solving them. The equation $x^2 = 4$ is the relationship $y = x^2$ where y = 4. To solve this equation, it is obvious that we need only apply the properties of inverse operations and equality to solve it by taking the square root of both sides of the equation,

$$\sqrt{x^2} = \sqrt{4}.$$

Here we must be cautious, however, for although we know that 2^2 will yield 4, we must not forget that $(-2)^2$ will yield the same. Since we don't know whether the original variable was a positive 2 or a negative 2, we must allow for the possibility of either so the inverse operation on the above equation must yield two answers,

$$x = 2 \quad \text{and} \quad x = -2.$$

We often write this as $x = \pm 2$.

Whenever the equation contains only an x^2 term and a constant, the square root method may be applied directly. As stated before there will be one, two, or no solutions depending on the location and identity of the parabola. The two examples below illustrate this.

Example

Solve the quadratic equation $x^2 = 0$.

We know that for the parabola $y = x^2$, the location where y = 0 contains the vertex so we expect only one solution and that is, indeed, what we get,

$$x^2 = 0$$

$$\sqrt{x^2} = \sqrt{0} \quad \text{yields} \quad x = 0.$$

Example

Solve $-4x^2 + 3 = 4$ for x.

If we draw a quick sketch of the graph of $y = -4x^2 + 3$ (as done below) we will see that the parabola opens downward and there is no parabola at the line y = 4. Therefore, we do not expect a solution to this equation.

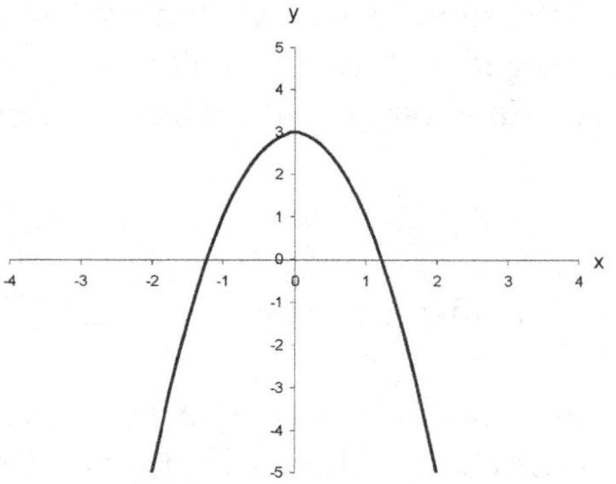

Rearranging the equation to obtain a single quadratic variable and constant and applying the square root method, we obtain

$$-4x^2+3=4$$

$$-4x^2=1$$

$$x^2=-\frac{1}{4}$$

$$\sqrt{x^2}=\sqrt{-\frac{1}{4}}.$$

Here, we must realize that $\sqrt{-\frac{1}{4}}$ does not exist confirming that there are no solutions for this equation.

8.1.1 Puzzling it All Out

Learning to play a musical instrument, build a canoe, or any other worthwhile endeavor requires active involvement before the information provided can be fully integrated. The study of mathematics is no different.

Work each of these puzzles on your own until you are sure you have the correct answer. Then check the answer in the back of the book. If you are having trouble agreeing with the book's answer you can: check the solutions manual if one is available in your area or ask an online program to show you a solution, ask a friend or a tutor, or ask your teacher for advice. Keep in mind that there may be more than one way to solve each problem. Always look for the easiest solution.

It is preferable that you understand how to get the correct solution for every puzzle on this page before moving to the next section.

1. Solve $\frac{2}{3}x^2 - 2 = 4$ for x.

2. Solve $\frac{2}{3}r^2 - 2 = -4$ for r.

3. Use Pythagorean's Theorem, $c^2 = a^2 + b^2$, where a and b are the two perpendicular sides and c is the longest side (hypotenuse) of a right triangle, to find the missing side,

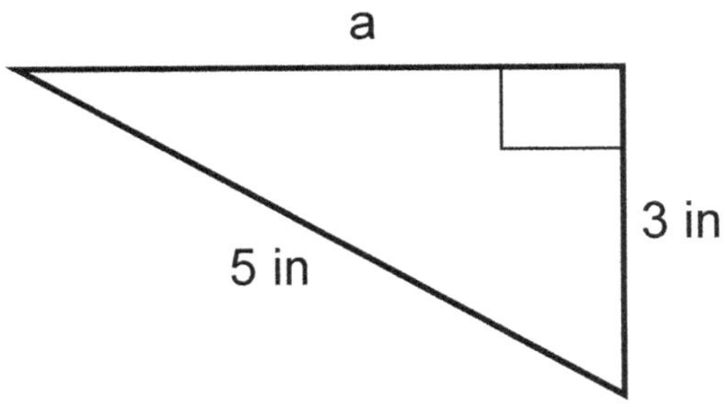

Why are we not concerned with the negative solution in this problem?

8.2 – Completing the Square (Extension of the Square Root Method)

8.2.1 – Standard Form of a Quadratic Equation

If we rearrange the vertex-curvature formula we found in Unit 2, we will see that the most general form of a quadratic equation contains a linear component,

Beginning with the vertex-curvature formula	$y = a(x+b)^2 + c$
Applying the definition of exponents	$y = a(x+b)(x+b) + c$
Applying the distribution method twice (FOIL first)	$y = ax^2 + axb + axb + ab^2 + c$
Combining like terms and constants	$y = ax^2 + (2ab)x + (ab^2 + c)$

gives,

$$y = ax^2 + bx + c$$

General Form for a Quadratic Equation
where x and y are any real numbers, and a, b, and c are constants specific to the equation.

8.2.2 – How Recognizing a Perfect Square can Save the Day

An equation of this form cannot readily be solved using the square root method as it is impossible to isolate one squared variable on one side of the equal sign and only a constant on the other. There are, however, some tricks we can use to modify the equation so the method will work. If we recognize a perfect square in the equation we can remedy the situation forthright,

Example

Solve $16 = x^2 + 2x + 1$ *for x.*

We should immediately recognize that this equation is really $16 = (x+1)^2$.

We can now apply the square root method, taking the square root of each side,

$\sqrt{16} = \sqrt{(x+1)^2}$ yielding $4 = x + 1$ or $-4 = x + 1$ so

$x = 3$ or $x = -5$.

8.2.3 – How to Make a Perfect Square

Of course, very few quadratic equations are perfect squares so the method in Section 9.2.2 would appear to have quite limited application. A closer look at multiplication of binomials through the Distributive Property when only one variable is involved, however, suggests that a perfect square may be constructed from any trinomial (squared term, linear term, and constant) by the addition or subtraction of a constant as shown below.

Example

Solve the quadratic equation $11 = m^2 + 6m + 4$ *for m.*

This time there is no perfect square on the right hand side. However, if we compare this to the general equation for a perfect square,

$$(x+a)^2 = x^2 + 2ax + a^2$$

where x is a variable real number and a is a constant particular to the equation,

we will notice that we would have a perfect square if the constant were 9 instead of 4 as

$$(m+3)^2 = m^2 + 2(3)x + 3^2 = m^2 + 6m + 9.$$

Since we have an equation, not just an expression, we can change the 4 to a 9 on the right hand side by adding 5 to both sides of the equation,

$$11 + 5 = m^2 + 6m + 4 + 5$$

$$16 = m^2 + 6m + 9$$

$$16 = (m+3)^2$$

$$\sqrt{16} = \sqrt{(m+3)^2}$$

$m + 3 = 4$ or $m + 3 = -4$

$m = 1$ or $m = -7$.

8.2.4 Puzzling it All Out

Learning to play a musical instrument, build a canoe, or any other worthwhile endeavor requires active involvement before the information provided can be fully integrated. The study of mathematics is no different.

Work each of these puzzles on your own until you are sure you have the correct answer. Then check the answer in the back of the book. If you are having trouble agreeing with the book's answer you can: check the solutions manual if one is available in your area or ask an online program to show you a solution, ask a friend or a tutor, or ask your teacher for advice. Keep in mind that there may be more than one way to solve each problem. Always look for the easiest solution.

It is preferable that you understand how to get the correct solution for every puzzle on this page before moving to the next section.

1. Solve $x^2 - 8x + 16 = 25$ for x.

2. Solve $3r^2 + 6r + 3 = 12$ for r.

3. Solve $p(p + 10) + 13 = 4$ for p.

4. Solve $x^2 + bx = c$ for x.

8.3 – Quadratic Formula (General Expression of Completing the Square)

Completing the square provides a tool that can solve any quadratic equation. We might notice however, after performing this method on several different equations, that the method is identical regardless of the problem. Mathematicians in the past, always looking for ways to make repetitive problem solving easier, noticed this and developed the Quadratic Formula – a formula that can solve any quadratic equation once it is the standard form.

Class Activity

Solve the equation $ax^2 + bx + c = 0$ for x.

[Hint – Apply the Completing the Square Method (like you did in 9.2.4 Puzzle 4).]

If you don't have a class, try this for yourself using what we have learned so far.

The step by step solution to this Activity **(the derivation of the Quadratic Formula)** can be found in the Answers Section under 9.3 Class Activity.

We find from our derivation that any quadratic equation can be solved using the

Quadratic Formula

$$x = \frac{-b \pm \sqrt{b^2 - 4ac}}{2a}$$

where x = the specific answer(s) to a particular quadratic equation and a, b, and c, are the constant coefficients when the equation is in the standard form, $ax^2 + bx + c = 0$.

The quadratic formula may be used to solve any single variable quadratic equation. It is often convenient to use this formula on equations that do not readily lend themselves to other methods. The method is straightforward. Each of the constants is substituted into the formula and the resulting expression is simplified.

Example

Solve $4x^2 + 2x - 6 = 0$ *for x*

Although we could complete the square to solve this equation, it is already in the standard format and we can use the Quadratic Formula directly. We first identify each of the constants,

a = 4, b = 2, c = -6 .

We can then substitute the constants directly into the Quadratic Formula,

$$x = \frac{-2 \pm \sqrt{2^2 - 4(4)(-6)}}{2(2)}$$

$$x = -\frac{1}{2} \pm \sqrt{100}$$

$$x = -\frac{21}{2} \quad \text{or} \quad x = \frac{19}{2}.$$

8.3.1 – The Determinant

We note that only the quantity under the radical (square root sign) can produce two solutions. If this quantity is zero, there is only one solution, $x = \frac{-b}{2a}$. If the quantity under the radical is negative, there are no solutions. This quantity is often referred to as the Determinant of the Quadratic Equation since its value determines the number of solutions available in a given equation. We can summarize the meaning of the

Sign of the Determinant as	
Positive	Two solutions
Zero	One Solution
Negative	No solutions.

With practice, many equations with no solutions can be identified through the Determinant without going through the work of actually solving them.

8.3.2 Puzzling it All Out

Learning to play a musical instrument, build a canoe, or any other worthwhile endeavor requires active involvement before the information provided can be fully integrated. The study of mathematics is no different.

Work each of these puzzles on your own until you are sure you have the correct answer. Then check the answer in the back of the book. If you are having trouble agreeing with the book's answer you can: check the solutions manual if one is available in your area or ask an online program to show you a solution, ask a friend or a tutor, or ask your teacher for advice. Keep in mind that there may be more than one way to solve each problem. Always look for the easiest solution.

It is preferable that you understand how to get the correct solution for every puzzle on this page before moving to the next section.

1. Determine the number of solutions for $3x^2 - 2x + 16 = 0$.

2. Determine the number of solutions for $3x^2 - 6x + 3 = 0$.

3. Solve $\frac{4}{x} + 2 = 3x$ for x.

8.4 – Factoring

Although we can solve any quadratic equation with the Quadratic Formula, a simpler, and quicker, solution is often available through factoring. As we have seen before, factoring involves dividing a quantity into its various factors (multipliers). We have used this on whole numbers to find common divisors or multipliers to remove rational forms.

We are also factoring when we use the Distributive Property in reverse as we did in Unit 1. When working forward we multiply a factor through an associated quantity of terms,

$$3(x^2 + 2x - 4) = 3x^2 + 6x - 12.$$

When working backward, we find a common factor in a series of terms and rewrite the expression as a combination of two factors,

$$3x^2 + 6x - 12 = 3(x^2) + 3(2x) - 3(4) = 3(x^2 + 2x - 4).$$

We have also seen how the multiplication of two binomials often results in a trinomial,

$$(2x - 4)(x + 3) = 2x^2 + 6x - 4x - 12 = 2x^2 + 2x - 12.$$

Factoring can, therefore, also be performed on many trinomials to return them to the two binomial factors that produced them. Once we have our equation in the form

$$(ax - b)(ex + f) = 0,$$

we can use a well known property of multiplication to obtain a direct solution.

8.4.1 – Application of the Zero Product Property

We all know that if we have something zero number of times, we don't have anything. Anything times 0 is equal to zero or, expressed mathematically,

$$a(0) = 0.$$

From this it is also possible to see that if we are multiplying two things and we get zero as a product (answer), one of the things we are multiplying must be equal to zero or, expressed mathematically,

If $ab = 0$

Then either

$a = 0$ or $b = 0$

Zero Product Property.

Application of the Zero Product Property to a quadratic equation containing the trinomial developed earlier, $(2x-4)(x+3)=2x^2+2x-12$, will illustrate the method.

Example

Solve the equation $2x^2+2x-12=0$ *for x.*

As we have already seen, the trinomial on the left is the product of two binomials and we can readily substitute.

$$(2x-4)(x+3)=0.$$

Applying the Zero Product Property directly we find that

$$(2x-4)=0 \quad \text{yielding} \quad x=2$$

or

$$(x+3)=0 \quad \text{yielding} \quad x=-3.$$

8.4.2 – Factoring Methods

Finding two binomials whose product produces a trinomial we need to factor can be challenging but, with practice, factoring and use of the Zero Product Principle can provide a much simpler and more direct solution than the tedious substitution and evaluation required when using the Quadratic Formula.

8.4.2.1 – Factoring $x^2 + bx + c$ forms

The most basic factoring method, and the one most often used, we will call educated guessing. Basically, we use what we know about multiplying two binomials to "unmultiply" them. To illustrate this we must first review the enhanced distributive property, or the FOIL method, for multiplication of binomials. Let us take the general case of $(x + a)(x + b)$.

When we multiply the first two terms we will get x^2. This is the only multiplication that will give a squared term so we know that the squared term in the resulting trinomial comes from the multiplication of the first two terms. The constant term in the trinomial is equally easy to trace as the only terms in the binomial that will produce a constant when multiplied are the last two or a times b. The middle two multiplications, the inner and the outer, each produce a term with a linear variable, in this case ax and bx. To complete the multiplication these terms are added and their sum is the linear term in the trinomial. To summarize:

Any trinomial of the form

$$x^2 + bx + c$$

(where x is a variable we are studying and a, b, and c are particular constants)

can be factored into a binomial of the form

$$(x + e)(x + f)$$

(where x is a variable we are studying and e and f are particular constants).

where $e(f) = c$, and $b = e + f$.

We can factor a trinomial of the form $x^2 + bx + c$, by finding two constants whose product is c and whose sum is b. We will illustrate the method with an example.

Example

Solve the equation $x^2 + 5x + 6 = 0$

We first look for the factors of 6. The prime factors are 3, 2, and 1. We need to find a combination of these that when added together will give us the 5 in 5x. Obviously, 3 and 2 add to five so $3(1) = 3$ and 2 are the two constants we are looking for and we get

$$(x + 3)(x + 2) = x^2 + 5x + 6.$$

Multiplying the binomials provides a check that we are correct.

Once we have factored the trinomial we can substitute the binomials and use the Zero Product Property to solve the equation.

$$(x + 3)(x + 2) = 0$$

$(x + 3) = 0$ yields $x = -3$

or

$(x + 2) = 0$ yields $x = -2$.

Food for Thought

How is the method affected when there is a negative sign in the trinomial?

8.4.2.2 – Factoring $ax^2 + bx + c$ forms

One little constant can make a lot of trouble. To understand why, we need to understand where this form comes from. In our previous analysis we discovered that the squared term in a trinomial must come from the product of the first two terms of the multiplication of binomials. The only way this term can have a constant is if one or more of these two terms has a constant. The form

$$ax^2 + bx + c$$

(where x is a variable we are studying and a, b, and c are particular constants)

will always come from a binomial of the form

$$(mx + e)(nx + f)$$

(where x is a variable we are studying and m, n, e, and f are particular constants).

The addition of these constants complicates the factoring as the l and m constants end up in every term but the last when multiplied as seen below:

$$(mx + e)(nx + f) = lmx^2 + mfx + nex + ef = lmx^2 + (mf + ne)x + ef\ .$$

Obviously guessing will not be quite as effective in this case.

8.4.2.3 – The ac Method of Factoring

A suitable method has been developed to factor trinomials of the $ax^2 + bx + c$ form called the ac method. The trick comes from the fact that the middle term is a sum of factors of the product of a and c. We will summarize this method in the following example.

Example

Solve the equation $4x^2 - x - 3 = 0$ *for x.*

First, we identify the constants,

a = 4, b = -1, and c = -3.

To apply the ac method we first multiply the constant a by the constant c, 4(3) = -12.

We then look for pairs of numbers that will produce this product. In our case there are

-1 and 12 or 1 and -12

-6 and 2 or 6 and -2

and

-3 and 4 or 3 and -4.

Next, we look for two of these numbers that will add to the constant in the middle, (-1).

We find that -4 and 3 work nicely.

Now, we can use these two constants to rewrite the initial equation

$$4x^2 - 4x + 3x - 3 = 0.$$

Regrouping helps us collect like terms

$$(4x^2 - 4x) + (3x - 3) = 0$$

We can then factor out like terms using the distributive property backwards

$$4x(x-1) + 3(x-1) = 0.$$

The common factor (x – 1) allows us to use the backward distributive property again.

Collecting the common factors results in $(4x+3)(x-1) = 0$.

We can then apply the Zero Product Property to solve the equation

$4x + 3 = 0$ yields $x = -\frac{3}{4}$ or

$x - 1 = 0$ yields x = 1.

8.4.3 – Recognizing Predictable Forms

Of course, if the trinomial is already in a form that the generating binomial is easily recognizable, our work is simple. When we can recognize forms such as Perfect Squares, the factoring method can be very quick and easy.

Example

Solve $x^2 - 14x + 49 = 0$ for x

We recognize this as the perfect square $(x - 7)^2$ and make the substitution.

$$(x-7)^2 = 0 \text{ or } (x-7)(x-7) = 0.$$

We can either take the square root of each side or use the Zero Product Property to find

x = 7.

8.4.4 Puzzling it All Out

Learning to play a musical instrument, build a canoe, or any other worthwhile endeavor requires active involvement before the information provided can be fully integrated. The study of mathematics is no different.

Work each of these puzzles on your own until you are sure you have the correct answer. Then check the answer in the back of the book. If you are having trouble agreeing with the book's answer you can: check the solutions manual if one is available in your area or ask an online program to show you a solution, ask a friend or a tutor, or ask your teacher for advice. Keep in mind that there may be more than one way to solve each problem. Always look for the easiest solution.

It is preferable that you understand how to get the correct solution for every puzzle on this page before moving to the next section.

1. Solve the equation $x^2 + 2x + 4 = 0$.
2. Solve the equation $x^2 - 12x + 36 = 0$.
3. Solve $x^2 - 3x = -2$ for x.
4. Solve $q^2 + \frac{5}{2}x = \frac{3}{2}$ for q.

8.5 – Using Quadratic Equations to Solve Problems

Many real life problems translate into quadratic equations. We will look at only two of these in this section.

Example

The City wants to add a square garden split down the middle with a 10' path between for access and seating to a public park. The park has a very open geometry, but the total garden area (including the path) can be no larger than $\frac{1}{2}$ acre. How long should each side be if they want to keep the shape of the garden before splitting as a square and use the maximum allowable acreage?

Like all geometry problems, a diagram will help us translate.

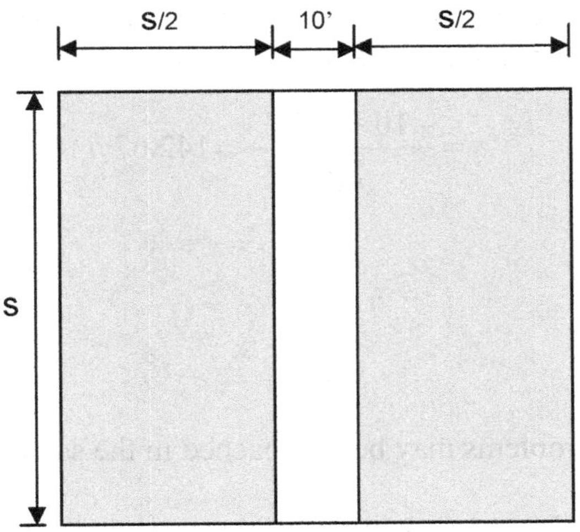

We can then apply applicable geometry formulas. In this case, we will need both the formula for the area of a square, ($A = s^2$) where A is the area and s is the length of one side of the square, and of a rectangle, ($A = lw$) where A is the area of the rectangle and l and w are the dimensions of the length and width respectively. We know only the width of the rectangle, but we also know that its length is equal to s. We also know that the sum of the two areas (one square and one rectangle) must add to $\frac{1}{2}$ acre. Making the appropriate substitutions we get:

$$\frac{1}{2}acre = s^2 + 10ft(s).$$

The first thing we notice is that there are units of both acres and feet in the problem and that may cause us trouble. If we know (or look up) that 1 acre = 43,560 ft^2 we can make the substitution,

$$\frac{1}{2}\left(43560\,ft^2\right) = s^2 + 10\,ft(s).$$

Since the units are all the same, we can drop them from the equation knowing that the answer will be in feet. Simplifying and moving to standard format we obtain,

$$s^2 + 10s - 21780 = 0.$$

The equation does not look easy to factor and is not a recognizable form so we are left with completing the square or using the Quadratic Formula as options for solution. We will use the Quadratic Formula here. First we identify the constants, then make the substitution:

a = 1, b = 10, c = -21780.

$$s = \frac{-b \pm \sqrt{b^2 - 4ac}}{2a} = \frac{-10 \pm \sqrt{10^2 - 4(1)(-21780)}}{2(1)}.$$

Although the square root yields both 295.33 and -295.33 as possible solutions, it is easy to see that the negative number will give us a negative (and therefore meaningless) result for the dimension of the side so we will use only the positive.

$$s = \frac{-10 + 295.33}{2(1)} = 142.67\,ft.$$

Many other geometric problems may be approached in the same way.

As we have seen in Unit 2, many scientific, business, and other practical formulas result in quadratic equations. We must often solve these to obtain the information we need. One common problem, shown in the next example, illustrates the almost magical power of mathematics in allowing us to solve complicated problems with ease.

Example

Dan unthinkingly shoots his shot gun into the air while standing in a crowd of friends. The shot is catapulted into the air at a velocity (speed) of 1200 feet pre second. Using the formula for uniform acceleration of an object,

$$y = v_0 t + \frac{1}{2}at^2$$

(where y = the total displacement of the object, v_0 = the initial velocity of the object, a = the acceleration of the object, and t = total time the object travels),

to calculate how much time his friends have to get out of the way before it begins raining buckshot.

The solution to this problem involves a trick. We must realize that if the gunshot first goes up and then, the actual total displacement is zero (it ends up where it starts). The only acceleration in this problem is that of gravity (32 ft/sec^2). Since the object moves up while gravity is pulling down, the acceleration is working against the movement and is negative. We can then make the appropriate substitutions into the formula. Since all the units are in feet and seconds, we can leave them out of the equation for clarity. We obtain

$$0 = 1200t + \frac{1}{2}(-32)t^2 .$$

We can factor out the common t easily

$$0 = \left(1200 + \frac{1}{2}(-32)t\right)t .$$

Applying the Zero Product Property we get

t = 0

(The preferable case where the shotgun is never fired)

or

$$1200 + \frac{1}{2}(-32)t = 0 \quad \text{yielding} \quad t = \frac{1200}{16} = 75\,\text{sec}$$

t = 75 sec = 1 minute and 15 seconds - time to start running!

[Note – The buckshot will return at the same velocity (speed) it started.]

8.5.1 Puzzling it All Out

Learning to play a musical instrument, build a canoe, or any other worthwhile endeavor requires active involvement before the information provided can be fully integrated. The study of mathematics is no different.

Work each of these puzzles on your own until you are sure you have the correct answer. Then check the answer in the back of the book. If you are having trouble agreeing with the book's answer you can: check the solutions manual if one is available in your area or ask an online program to show you a solution, ask a friend or a tutor, or ask your teacher for advice. Keep in mind that there may be more than one way to solve each problem. Always look for the easiest solution.

It is preferable that you understand how to get the correct solution for every puzzle on this page before moving to the next section.

1. The University wants to add a brick border around their back patio. The patio currently measures 200' by 200'. They have only 48,400 total square feet to use. How wide should the border be if they want it to be of uniform width completely surrounding the patio?

2. Use the formula

$$x = v_0 t + \frac{1}{2}at^2$$

(where x = the displacement (or distance traveled), v_0 = the initial velocity an object is traveling, and a is the acceleration of the object

to determine how long it will take a rocket that breaks the earth's gravity with an initial speed of 1500 m/s (meters per second) and an acceleration of 20 m/s^2 (meters per second squared) to reach the moon, a distance of approximately 400 million meters?

8.6 Chapter 8 – Review

Quadratic relationships are relationships that have a variable with an exponent of 2. They may be represented by the

Standard Vertex-Curvature form for a Quadratic Equation,

$$y = a(x-b)^2 + c$$

where x and y are any real numbers, the point (b,c) is the vertex of the parabola, and a is a constant that modifies the curvature of the parabola.

Unlike a linear relationship that always has one unique value of x for each value of y; quadratic relationships have regions where each value of x can return none, one, or two values for y as shown below.

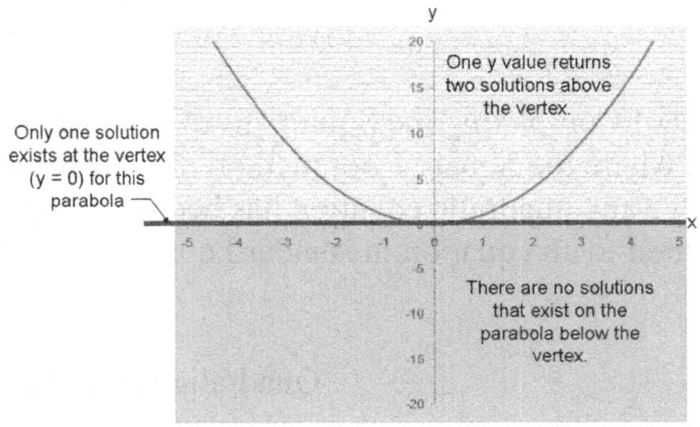

We often have quadratic equations that represent a particular parabola at a particular value of y and need to be solved for the corresponding x value. These equations come in many forms but can always be reduced to the

Standard Form for a Quadratic Equation,

$$ax^2 + bx + c = 0$$

where x and y are any real numbers, and
a, b, and c are constants particular to the parabola.

There are three main ways to solve quadratic equations; the Square Root Method (with its extension Completing the Square), the Quadratic Formula, and the use of Factoring and the Zero Product Property.

The Square Root method is the simplest method but can only be applied when there is a perfect square in the variable and a constant on the other side. If we can get the equation into this form, we have but to take the square root of each side to obtain the answer as,

$$x^2 = \frac{4}{25} \text{ yields}$$
$$\sqrt{x^2} = \sqrt{\frac{4}{25}}$$
$$\text{giving } x = \frac{2}{5} \text{ or } x = -\frac{2}{5},$$

or

$$(x-3)^2 = 16 \text{ yields}$$
$$\sqrt{(x-3)^2} = \sqrt{16}$$
$$\text{giving } x = 7 \text{ or } x = -1.$$

The method of Completing the Square gives us a way to convert any quadratic equation into a form where the Square Root method may be used. Accordingly, a general formula that can solve any quadratic equation has been developed by applying the Completing the Square method to an equation in standard quadratic form. The resulting

Quadratic Formula,

$$x = \frac{-b \pm \sqrt{b^2 - 4ac}}{2a}$$

where x = the specific answer(s) to a particular quadratic equation and a, b, and c, are the constant coefficients when the equation is in the standard form,

can be used to solve any quadratic equation but can often be tedious to apply.

The quantity under the radical, often called the Determinant, determines the number of solutions available in a given equation. We can summarize the meaning of the of the

Sign of the Determinant

Positive	Two solutions
Zero	One Solution
Negative	No solutions.

A more direct, and often faster and easier, solution is available if we can factor the trinomial on the left side of a standard equation into two binomial linear factors. Then the

Zero Product Property,

If $ab = 0$, Then either $a = 0$ or $b = 0$

may be used to obtain an answer directly. Factoring of trinomials can be done by recognizing common forms such as perfect squares, educated guessing (for $x^2 + bx + c$ forms), or the ac method (for $ax^2 + bx + c$ forms).

Practice with the Distributive Property, both forward and backward, can help a person become more comfortable with factoring trinomials.

8.6.1 – More Practice Puzzles

1. Solve $x^2 - 4x + 3 = 0$ for x.

2. Solve the equation $\dfrac{t^2 - 2t - 8}{t + 2} = 0$ for t.

3. Solve the equation $\dfrac{1}{3}x^2 - \dfrac{2x^2}{3} + \dfrac{1}{6}x = \dfrac{2x}{3}$ for x.

4. Solve the equation $x^2 - (a + c)x + ac = 0$ for x.

5. What are the coordinates of the Point P?

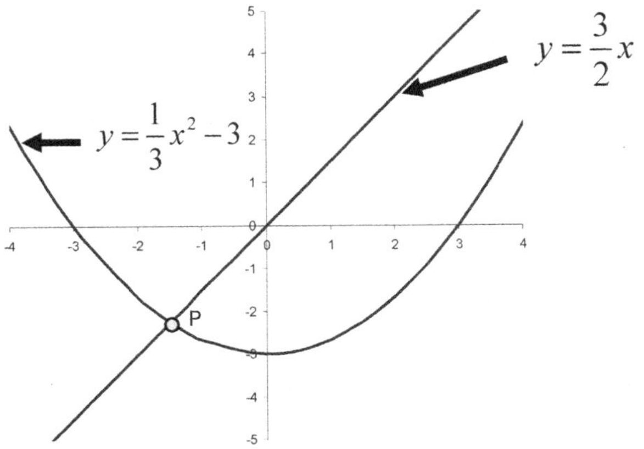

6. Solve the equation $p^2 - 4t^2 = 0$ for p.

Chapter 9 – Bringing it All Together

In this book we have looked at how we use letters and other symbols to represent variables in Mathematics. We have looked at the three major reasons for using variables;

1. To Make General Mathematical Statements,
2. To Explore Relationships, and
3. To Solve for Unknowns.

We have seen that the mathematical properties applicable to numbers are also valid for letters, or other variables. The use of variables allows us to express these properties as general statements or rules. We have found that there are only a few properties that govern all the rules for arithmetic. We wrote these properties for the basic arithmetic operations and their inverses; addition and subtraction, multiplication and division, and exponents and roots. They are summarized below.

For all mathematical statements in this Chapter unless otherwise defined:
a, b, e, and f are equal to any real numbers

Addition and Subtraction

The most basic arithmetic operation is a shortcut for counting and uses the symbols + and -. We have written several properties for addition and subtraction including:

The Identity Property of Addition and Subtraction

$a + 0 = a$ or $a - 0 = a$

Commutative Property for Addition and Subtraction

$a + b = b + a$ or $a - b = (-b) + a$

Multiplication and Division

Multiplication, as well as it's inverse, division, is a shortcut for addition and subtraction. We symbolize multiplication by writing one number or letter next to each other. We often separate these using parentheses or a small dot for clarity. Division is represented by writing the number being divided above the number of groups it is being divided into separated by a line. Multiplication can be expressed somewhat generally defined using variables as:

Definition of Multiplication

$$3a = a + a + a$$

but requires series notation, not covered in this class, for a completely general expression.

Division is defined as the inverse of multiplication:

$$\frac{ab}{b} = a$$

We wrote several properties for multiplication and division as shown below.

Identity Property for Multiplication and Division

$$a(1) = a \quad \text{or} \quad \frac{a}{1} = a$$

Commutative Property for Multiplication and Division

$$ab = ba \qquad \text{or} \qquad \frac{a}{e}\left(\frac{b}{f}\right) = \frac{a}{f}\left(\frac{b}{e}\right)$$

Zero Product Property for Multiplication

$$a(0) = 0$$

or

If $ab = 0$ Then either $a = 0$ or $b = 0$

Inversion of Divisor Property for Division by Rational Forms

$$\frac{\left(\frac{a}{b}\right)}{\left(\frac{e}{f}\right)} = \left(\frac{a}{b}\right)\left(\frac{f}{e}\right).$$

Exponents and Roots

We defined exponents as a shortcut for multiplication and again used a somewhat general expression to define them and their inverse, roots:

Definition of Exponential Format

$$a(a)(a)(a)(a)(a) = a^6$$

Inverse Operation for Exponents (Roots)

$$\sqrt[b]{(a^b)} = a$$

We were able to use the definition of exponents to develop a set of properties (rules) for exponents that can help us evaluate complex exponential expressions quickly and easily.

Rules for Exponents

If x, a, and b are any real numbers

Then $x^a(x^b) = x^{(a+b)}$

$$\frac{x^b}{x^a} = x^{(b-a)}$$

$$(x^a)^b = x^{(ab)}$$

$$x^1 = x$$

$$x^0 = 1 \quad \text{and,}$$

$$\sqrt[b]{x} = x^{\frac{1}{b}}$$

Order of Operations

We found that the symbology we use for higher order, or shortcut, operations can present problems when we combine them with lower order operations and that we could solve this problem by always performing higher order operations before lower order operations in an expression. We looked at the idea of Association, sometimes called the Associative Property, and learned that mathematicians commonly use parentheses to separate operations that should be performed together regardless of order and identified a common acronym, PEMDAS (Please Excuse My Dear Aunt Sally), to remind us to evaluate any Associated quantities (such as those in parentheses) first followed by Exponents, Multiplication, Division, Addition, and Subtraction respectively.

While we can always calculate expressions with only numbers using the Order of Operations, it is not always possible to do so when the expression contains variables. We used mathematical logic to show that it is sometimes possible to rearrange these expressions using the Distributive Property or the Expanded Distributive Property for binomials.

Distributive Property

$$a(b+e+f)=ab+ae+af$$

Expanded Distributive Property (FOIL)

$$(a+b)(c+f)=ac+af+bc+bf$$

We learned to recognize two special cases that helped us simplify many complex expressions quickly and easily,

Perfect Square Expansion

$$(a+b)^2=a^2+2ab+b^2$$

Difference of Squares

$$(a+b)(a-b)=a^2-b^2$$

Formulas

We found that by assigning letters to represent quantities that vary we can develop formulas that help us make repetitive calculations quickly and easily. Many formulas have been developed for routine geometric, scientific, business, and other problems.

We found that

$$P = 2(L + W) \quad \text{and} \quad A = LW$$

Where P = Perimeter, A = Area, L = Length and W = Width (for a rectangle),

$$A_T = \frac{1}{2}bh$$

where A_T = Area, b = base, and h = height (for a triangle), and

$$C = \pi d \quad \& \quad A = \pi r^2$$

where C = the circumference, A = the area, d = the diameter, and

$r = \frac{d}{2}$ or $\frac{1}{2}$ the diameter

(for a circle), and showed that

$$d = rt$$

where d = the distance covered, r = the rate (or speed) the object is traveling, and t = the time the object is moving at the given rate,

$$F = \frac{9}{5}C + 32$$

Where F = the temperature in Fahrenheit and C = the temperature in Celsius,

$$A = P(1 + rt)$$

where A = the total amount at the end of the investment, P = principle, r = rate, and t = time, and

$$C = F + vn$$

where C = Total Cost, F = Fixed Costs, v = variable cost per item, and n = number of items.

Relationships and the Cartesian Coordinate System

We looked at a way to draw pictures of mathematical relationships developed by Rene Descartes. When we draw a relationship in the Cartesian coordinate system, we can use it to compare with similar relationships, find specific values of the relationship at specific points in time, or make predictions based on the trends we see in our drawings. We defined this system as a set of three perpendicular axes intersecting at one point as shown below.

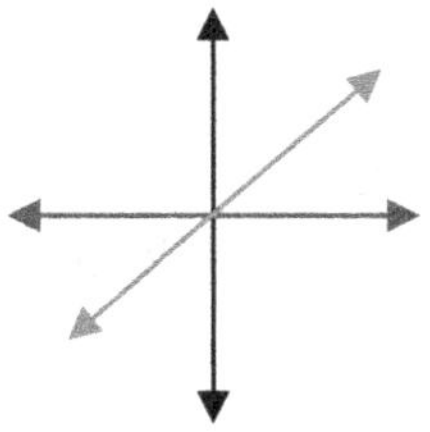

To illustrate the system we used just two of the axes to explore relationships with two variables. Each axis is divided and labeled uniformly with zero at the intersection and positive values getting larger moving up and to the right. Specific points may then be identified at the intersections of intervals as shown below.

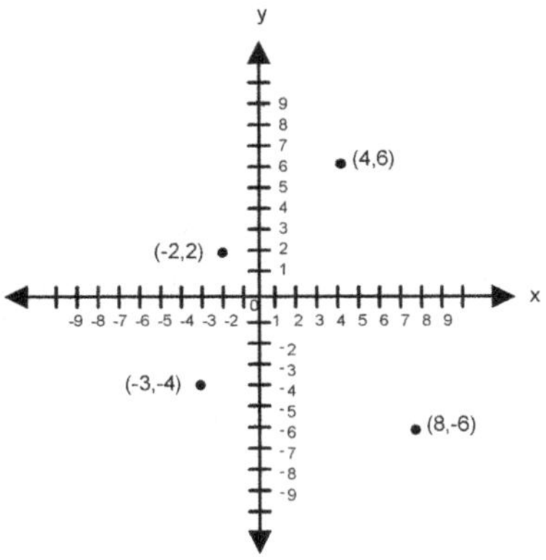

We found that by solving two variable equations for a variety of values and plotting the values on a set of Cartesian axes we could see similarities in many equations. Equations containing variables with an exponent of 1 always form a straight line. We called these relationships linear and found that any linear two variable relationship could be expressed in the form

$$y = mx + b$$

where x and y are any real numbers,

m = the slope of the line $\left(\frac{\Delta y}{\Delta x}\right)$, and b = the y value of the y intercept of the line.

We found that these lines had some common characteristics including:

Lines from equations with positive slopes slope right,
Lines from equations with negative slopes slope left,
Horizontal lines have a zero slope, only a y intercept, and follow the form y = b,
Vertical lines have no slope, only an x intercept, and follow the form x = a,

and

$m_1 = m_2$ for two Parallel Lines

$m_1 = -\frac{1}{m_2}$ for two Perpendicular Lines.

We found that relationships where one of the variables had an exponent of 2 formed a symmetric shape we called a parabola (shown below). We called this type of relationship quadratic.

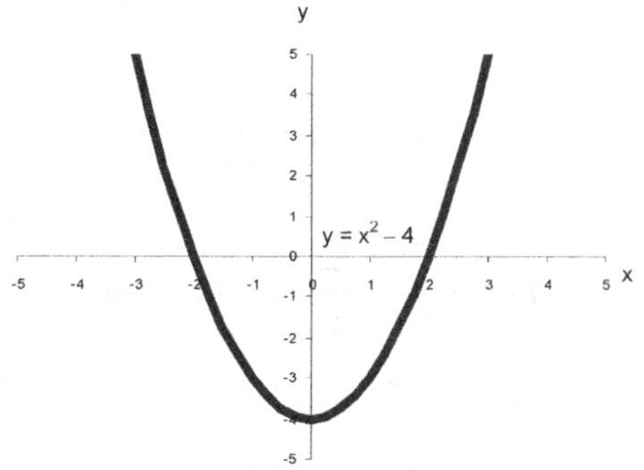

We found that, like linear relationships, we could express any quadratic relationship in the vertex-curvature form,

$$y = a(x-b)^2 + c$$

where the point (b,c) is the vertex of the parabola, and a modifies the curvature.

And observed that

Positive a results in a parabola opening upward,
Negative a results in a parabola opening downward.

Solving Equations

We then looked at the property of equality and used it, combined with the other tools we have discovered, to rearrange equations to provide more useful formulas or solve for unknowns. We found that we could unwind equations by using inverse operations in the reverse order of operations.

We used this with linear equations to rearrange to the y = mx + b format for easy graphing. We also symbolically rearranged this formula, using the definition for the slope, $y = \frac{\Delta y}{\Delta x} = \frac{y - y_1}{x - x_1}$, to obtain the

Point-Slope Formula for a line

$$m(x - x_1) = (y - y_1)$$

where m = the slope and (x_1,y_1) and (x_2,y_2) are points on the line,

giving us an easy way to write an equation from a line if we know the slope and at least one point.

We solved linear equations with one variable and found that, as we suspected from analyzing linear graphs, there is always only one solution. We found that equations having two linear variables required a set of two equations to solve. We can solve such a set of equations graphically, by finding their point of intersection, or algebraically using either the Method of Substitution, where one equation is solved for one of the variables and substituted into the other in that variables place, or the Method of Elimination, where one equation is added or subtracted from the other equation in a way that eliminates one of the variables.

We found that quadratic equations were often more difficult to solve and could have 1, 2, or no solutions. We saw that an equation that could be expressed as a squared term on one side with a constant on the other could be solved readily by taking the square root of both sides. Since any trinomial can be written as a perfect square of a binomial plus a constant, the method of Completing the Square has been developed. Completing the Square on the standard general form of a quadratic equation, $ax^2 + bx + c$, results in the

Quadratic Formula

$$x = \frac{-b \pm \sqrt{b^2 - 4ac}}{2a}$$

where x = the specific answer(s) to a particular quadratic equation and
a, b, and c, are the constant coefficients when the equation is in the standard form,

The quantity under the radical, called the Determinant, determines the number of solutions available in a given equation as:

Signs of the Determinant

Positive	Two solutions
Zero	One Solution
Negative	No solutions.

Often, it is quicker and easier to factor quadratic equations to obtain two terms being multiplied and equal to zero. We can then use the Zero Product Property to readily obtain solutions.

Review of Operations Involving Signs

Addition and Subtraction

Like Signs – Add and use the common sign

$2 + 4 = 6$ or $-3 - 5 = -8$

Unlike Signs – Subtract and use the sign of the larger

$-2 + 6 = 4$ or $3 - 8 = -5$

Multiplication and Division

Like Signs – Positive

$2(3) = 6$ or $-2(-3) = 6$

Unlike Signs – Negative

$\frac{-6}{3} = -2$ or $2(-3) = -6$

Exponents

Positive Base – Positive

$2^3 = 8$

Negative Base but Even Exponents – Positive

$(-2)^2 = 4$

Negative Base but Odd Exponents – Negative

$(-2)^3 = 8$

Roots

Positive Base – Positive

$\sqrt[3]{27} = 3$

Negative Base but Even Exponents – Does not Exist

$\sqrt{-4}$ or $\sqrt[8]{-256}$ cannot exist

Negative Base but Odd Exponents – Negative

$\sqrt[3]{-27} = -3$

Rules of Exponents

If x, a, and b are any real numbers then, for exponential operations:

To Multiply – Add Exponents

$$x^a(x^b) = x^{(a+b)}$$

To Divide – Subtract Exponents

$$\frac{x^b}{x^a} = x^{(b-a)}$$

To Raise to a Power – Multiply Exponents

$$\left(x^a\right)^b = x^{(ab)}$$

Raising to the Power on 1 Returns the Identity

$$x^1 = x$$

Raising to the Power on 0 Returns One

$x^0 = 1$ and,

A Root may be Written in Exponential Format:

$$\sqrt[b]{x} = x^{\frac{1}{b}}.$$

Answers to Problems

1.1.4

1. 50
2. 14
3. 2 – a
4. 3
5. H = Harry's Savings, J = Judy's Savings, S = the sum of Harry and Judy's Savings:

$$S = H + J$$

6. For a rectangle with the dimensions of its of length and width equal to a and b and Perimeter, P:

$$P = a + b + a + b$$

1.2.9

1. 24
2. 6
3. 1
4. 8
5. A = Amount in my piggy bank, w = number of weeks of savings

$$A = 5w$$

6. For a rectangle with sides lengths of a and b and an area of A,

$$A = lw$$

1.3.4

1. y^6
2. w^8
3. a^3
4. $(-2)^4 = 16$
5. $4 - 3$, $\frac{a}{a}$, and y^0 are three examples.
6. For x = any real number: $(-x)^2 = x^2$

1.4.7

1. $\frac{2}{6} + \frac{3}{6} = \frac{5}{6}$
2. $w^3 - 4w$

3. $\frac{yz(a^2)+xz(a^2)+2xy(a^2)}{xyz}=\frac{a^2(yz+xz+2xy)}{xyz}$
4. $f^{\left(\frac{3}{2}\right)}$
5. $\frac{xy(x+y)}{xy(y^2)}=\frac{x+y}{y^2}$
6. $\frac{(2x-3y)^2}{(2x-3y)}=2x-3y$
7. for a, x, b and y equal to any real numbers,

$$\frac{a}{x}+\frac{b}{y}=\frac{ay}{xy}+\frac{bx}{yx}=\frac{ay+bx}{xy}$$

1.5.4

1. 1
2. x
3. $\frac{y^2+9}{3y}$
4. $\frac{5}{2}$
5. $x^5 + 1$
6. 5

2.1.4

1. $P=19+\frac{9\pi}{2}\approx 33.14\cdot in.$
2. $A=\frac{3(4)}{2}+4(9)+\pi\left(\frac{9}{2}\right)^2=42+\frac{81}{4}\pi\approx 105.6\cdot sq.in.$
3. For a square with s = the length of one side, P = the perimeter, and A = the Area:

$$P=4s \qquad \text{and} \qquad A=s^2$$

The formulas are identical to a rectangle of sides of length a and b when $s = a + b$

2.2.3

1. $F=\frac{9}{5}(5)+32=41^{\circ}\text{F}$
2. $d_{today}=4\frac{miles}{hour}(15\cdot\min)\frac{1\cdot hour}{60\cdot\min}=4\left(\frac{1}{4}\right)=1\cdot mile$, $\frac{d_{today}}{d_{normal}}=\frac{1}{3}$

2.3.3

1. A = 100000[1 + 0.0025(120)] = \$130,000, the same.
   [Note – Monthly versus yearly returns can differ significantly when compound interest is considered.
2. \$1300

2.4.4

1. $A_6 = 6(6)^2 = 216$, $A_4 = 6(4)^2 = 96$, $\frac{A_6}{A_4} = \frac{216}{96} = \frac{3(8)(9)}{3(8)(4)} = \frac{9}{4}$
2. C = 20 + 5(120) = \$620
3. Let A = the amount saved after t (number of months), I = the initial deposit, and m = the monthly deposit, then:

   A = I + mt

   for John, I = \$100, m = \$50, and t = 24 months (after 2 yrs.) and,

   A = 100 + 50(24) = \$1300

   Yes, he will have the money for the \$320 mower.
4. $A = \frac{1}{2}bh = \frac{1}{2}(d)\left(\frac{d}{2}\right) = \frac{d^2}{4}$
5. Enrichment team – 9,235,220 organisms after 4 hours
   Starvation team – 12,970 organisms after 4 hours
6. $t = \frac{160}{40}$ = 4 hours. Yes, you (I) can make the trip.

3.2.4

1.

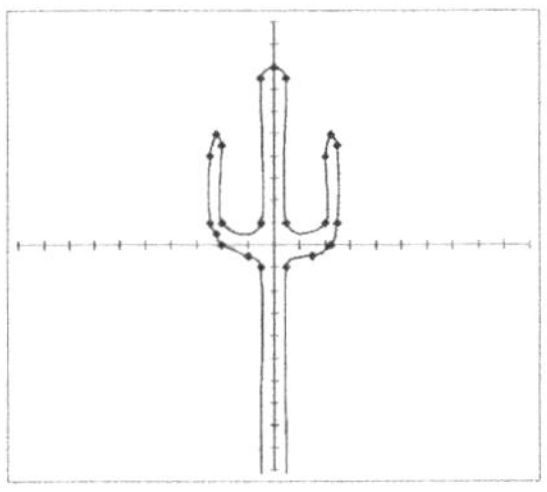

2.

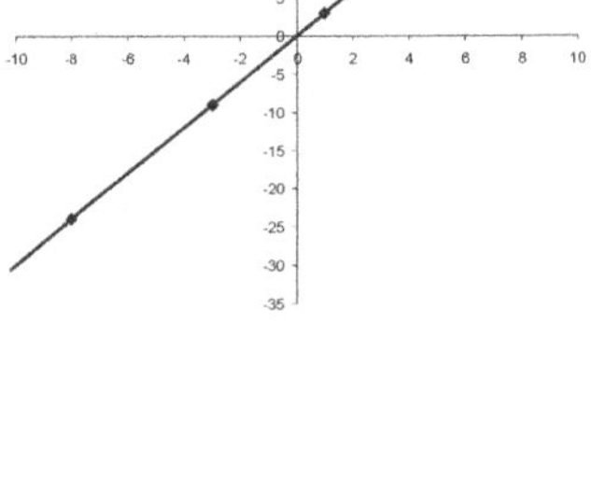

the dots form a straight line.

4.1.6

1. $c = 2\sqrt{19}$
2. 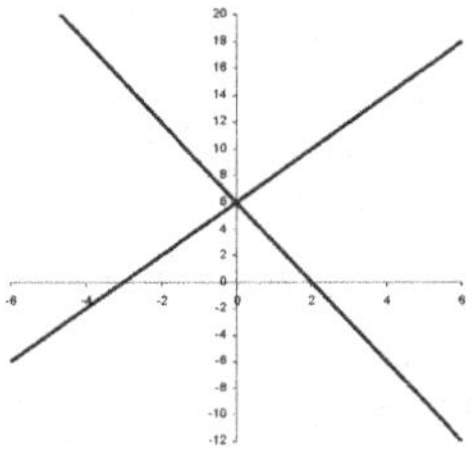
$A = \frac{1}{2}(5)(6) = 15$

4.2.5

1. (-6,0)
2. a) m = 4, y intercept – (0,-2), x intercept $\left(\frac{1}{2}, 0\right)$; b) non-existent slope, no y intercept, x intercept – (3, 0); c) m = 0, y intercept – (0,-5), no x intercept; d) m = r, y intercept – (0, d_o), x intercept - $\left(-\frac{d_o}{r}, 0\right)$
3. (6,-1)
4. Quadrant IV, The car will have a negative market value before I can pay off the loan.

4.3.1

1. 6 mos.
2. $m = -\frac{2}{3}$

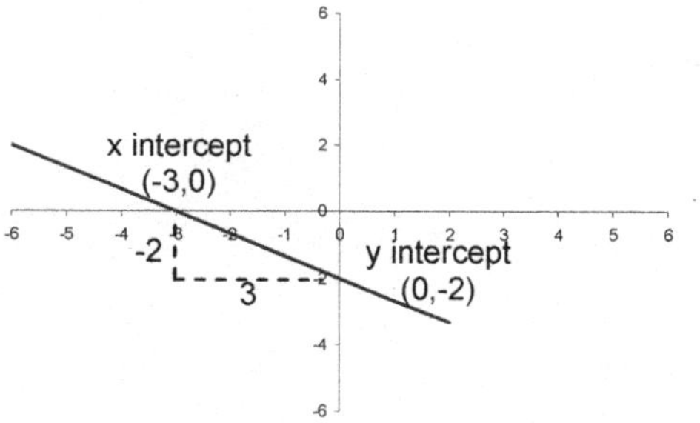

3. y = 4x – 3

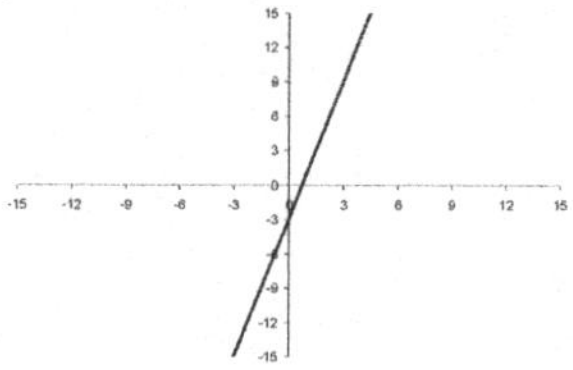

4. y = -4x, y = -4x – 2, y = -4x + 64
5. $y = \frac{1}{4}x$, $y = \frac{1}{4}x - 2$, $y = \frac{1}{4}x + 6$
6. (0,6)

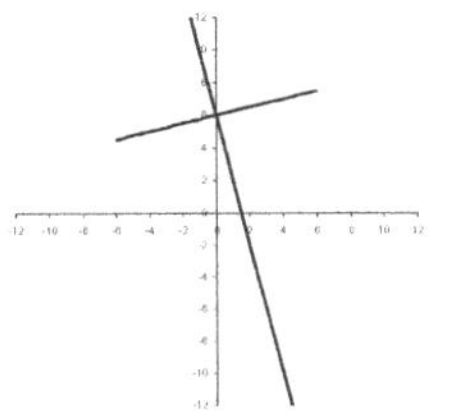

-6	30	4.5
-5	26	4.75
-4	22	5
-3	18	5.25
-2	14	5.5
-1	10	5.75
0	**6**	**6**

5.1.1

1.

x	y
0	-4
-1	-3
1	-3
-2	0
2	0
-3	5
3	5

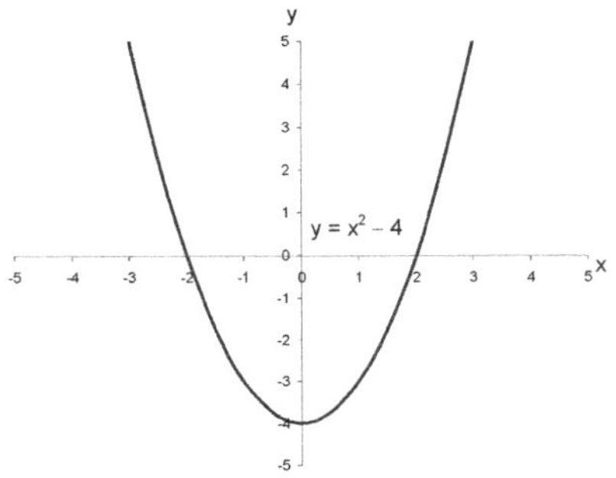

2.

x	y
0	-4
-1	-3
1	-3
-2	0
2	0
-3	5
3	5

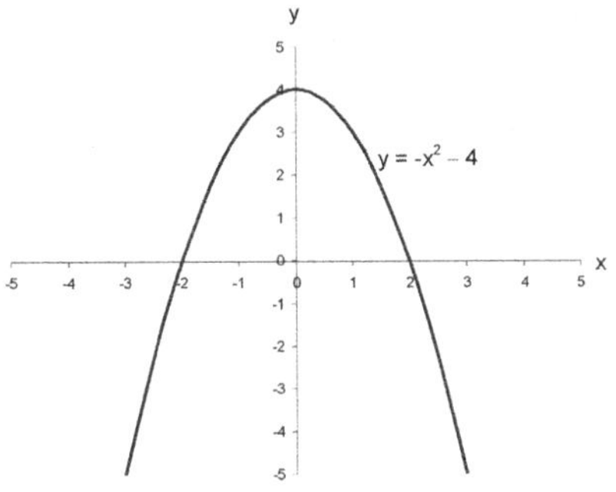

3. 1 and 5
4. 5

5.2.1

1. y = -x^2 + 2
2.

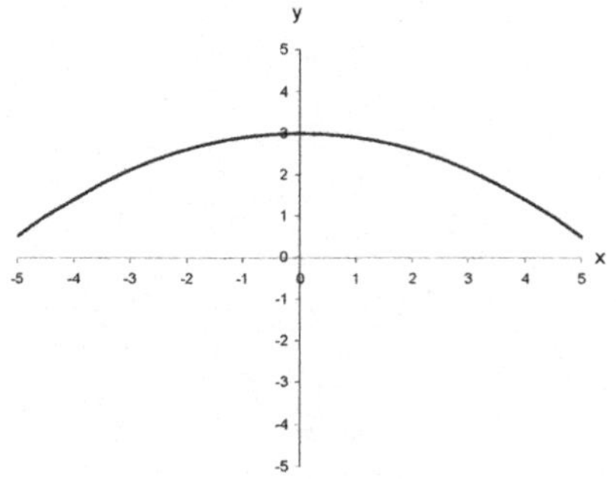

3.

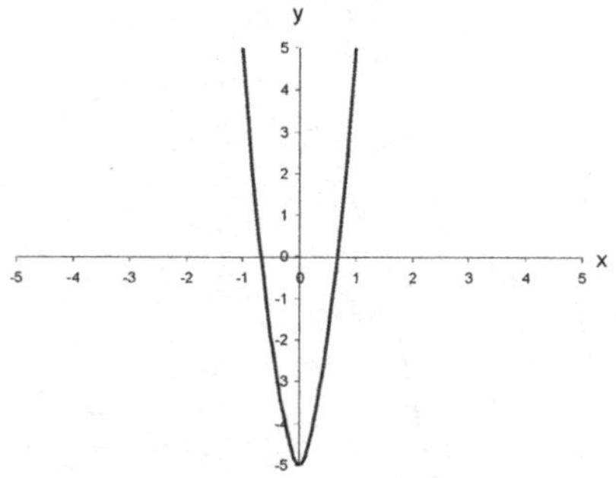

6.1.4

1. x = 6
2. y = 2 – x = -x + 2
3. x = 6
4. $r = \dfrac{d - d_0}{t}$
5. x = 8
6. y = 4

6.2.3

1. $b = \dfrac{2A}{h}$ where A is the area of the triangle, b is the length of one side of the triangle, and h is the length of a line running from a perpendicular to the base to the opposite vertex.
2. $r = \dfrac{A - P}{Pt}$ where r is the rate of investment, A is the total amount at the end of the investment, P is the initial principle invested, and t is the period of investment time.
3.

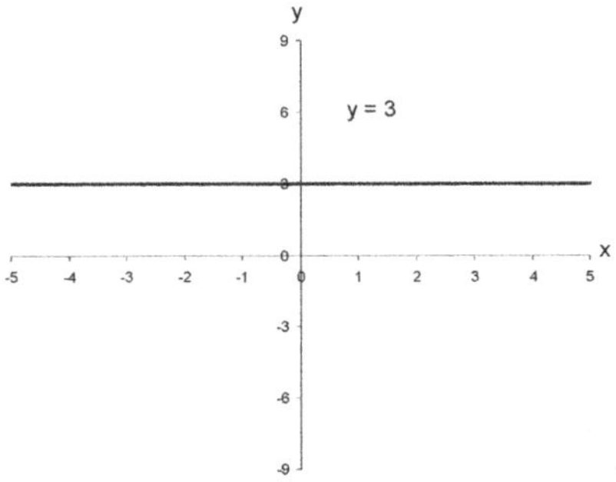

4. 2(x + 3) = y + 4 or y = 2x + 2, the line crosses x = 0 at y = 2.

6.3.1

1. x = 11
2. $r = \dfrac{2\gamma}{\rho g h}$ A taller tree requires a smaller tube radius.
3. $y = \dfrac{1}{5}x + 2$

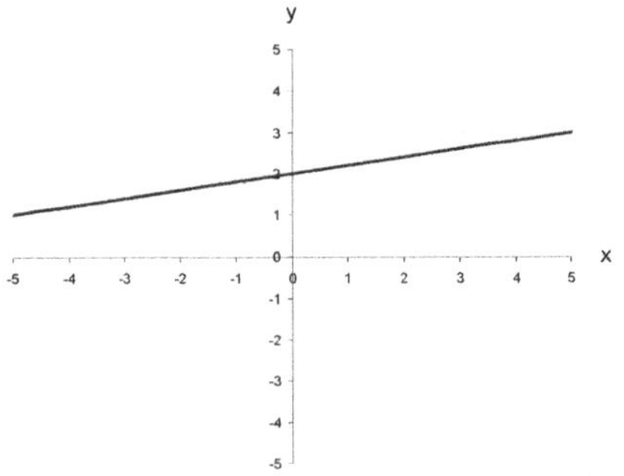

7.1.4

1. x = 15
2. x = 8
3. 12x + 1 = 325 – Each side should be 27’.
4. $\dfrac{5}{6}x + \dfrac{1}{6} = \dfrac{7}{2}$ - The material sheets must be 4’ long each.

7.2.3

1. d = 70t and d = 45t + 25 give t = 1 hr and d = 70 miles
2. p = 3, q = 2

8.3.1

1. y = 4x + 2
2. r = -3

3. $y = \frac{3}{4}x + 7$
4. x = 4
5. At the point (0,3)
6. c = the amount of cash in each envelope
   3c + \$125 = \$800
   There was \$225 in each envelope.

8.1.1

1. x = 3 or x = -3
2. There are no solutions.
3. The missing side is 4 in. We do not have to worry about the negative solution since a negative length of a triangle is meaningless.

8.2.4

1. x = 9 or x = -1
2. r = 1 or r = -3
3. p = -1 or p = -9
4. $x = -\frac{b}{2} \pm \sqrt{c + \left(\frac{b}{2}\right)^2}$ or, simplifying, $x = \frac{-b \pm \sqrt{b^2 - 4c}}{2}$

8.3 Class Activity

$ax^2 + bx + c = 0$ Given Equation

$ax^2 + bx = -c$ Subtract c from both sides

$\frac{ax^2 + bx}{a} = \frac{-c}{a}$ Divide both signs by a to remove quadratic coefficient

$x^2 + \frac{b}{a}x = \frac{-c}{a}$ Simplify. We now note that the expression on the left will be in the form of a perfect square $\left(x^2 + 2\left(\frac{b}{2a}\right)x + \left(\frac{b}{2a}\right)^2\right)$ if we add the constant term $\left(\frac{b}{2a}\right)^2$. (See next page)

$x^2 + \frac{b}{a}x + \left(\frac{b}{2a}\right)^2 = \frac{-c}{a} + \left(\frac{b}{2a}\right)^2$ Adding the term to both sides

$$x^2 + \frac{b}{a}x + \frac{b^2}{4a^2} = \frac{-c}{a} + \frac{b^2}{4a^2}$$ Distributing the exponent

$$\left(x + \frac{b}{2a}\right)^2 = \frac{-c}{a} + \frac{b^2}{4a^2}$$ Rewriting the left as a perfect square

$$\sqrt{\left(x + \frac{b}{2a}\right)^2} = \sqrt{\frac{-c}{a} + \frac{b^2}{4a^2}}$$ Taking the square root of each side

$$x + \frac{b}{2a} = \pm\sqrt{\frac{-c}{a} + \frac{b^2}{4a^2}}$$ Resolving the radicals

$$x = -\frac{b}{2a} \pm \sqrt{\frac{-c}{a} + \frac{b^2}{4a^2}}$$ Subtracting $\frac{b}{2a}$ from both sides.

The Formula is now usable, but the equation is somewhat complex.

$$x = -\frac{b}{2a} \pm \sqrt{\frac{-c}{a}\left(\frac{4a}{4a}\right) + \frac{b^2}{4a^2}}$$ Multiplying by one to obtain a common divisor

$$x = -\frac{b}{2a} \pm \sqrt{\frac{-4ac}{4a^2} + \frac{b^2}{4a^2}}$$ Simplifying

$$x = -\frac{b}{2a} \pm \sqrt{\frac{b^2 - 4ac}{4a^2}}$$ Combining

$$x = -\frac{b}{2a} \pm \frac{\sqrt{b^2 - 4ac}}{\sqrt{4a^2}}$$ Distributing the exponent

$$x = -\frac{b}{2a} \pm \frac{\sqrt{b^2 - 4ac}}{2a}$$ Resolving the radical

$$x = \frac{-b \pm \sqrt{b^2 - 4ac}}{2a}$$ Combining.

8.3.2

1. no solutions

2. one solution
3. $\frac{1}{3}\left(1 \pm \sqrt{13}\right)$

8.4.4

1. x = -2
2. x = 6
3. x = 2 or x = 1
4. $q = \frac{1}{2}$ or $q = -3$

8.5.1

1. The border should be 10' wide.
2. 6250 seconds or about 1.7 hours

8.6.1

1. x = 1 or x = -3
2. t = 4
3. $x = 0$ or $x = \frac{3}{2}$
4. x = a or x = c
5. $\left(-\frac{3}{2}, -\frac{9}{4}\right)$
6. $p = -2t$ or $p = 2t$

Index

Made in the USA
Monee, IL
08 November 2026

68020208R00103